Intelligent Systems Reference Library

Volume 123

Series editors

Janusz Kacprzyk, Polish Academy of Sciences, Warsaw, Poland
e-mail: kacprzyk@ibspan.waw.pl

Lakhmi C. Jain, University of Canberra, Canberra, Australia;
Bournemouth University, UK;
KES International, UK
e-mail: jainlc2002@yahoo.co.uk; jainlakhmi@gmail.com
URL: http://www.kesinternational.org/organisation.php

About this Series

The aim of this series is to publish a Reference Library, including novel advances and developments in all aspects of Intelligent Systems in an easily accessible and well structured form. The series includes reference works, handbooks, compendia, textbooks, well-structured monographs, dictionaries, and encyclopedias. It contains well integrated knowledge and current information in the field of Intelligent Systems. The series covers the theory, applications, and design methods of Intelligent Systems. Virtually all disciplines such as engineering, computer science, avionics, business, e-commerce, environment, healthcare, physics and life science are included.

More information about this series at http://www.springer.com/series/8578

Gregor Grambow · Roy Oberhauser
Manfred Reichert

Editors

Advances in Intelligent Process-Aware Information Systems

Concepts, Methods, and Technologies

 Springer

Editors
Gregor Grambow
Institute of Databases and Information
   Systems
University of Ulm
Ulm
Germany

Manfred Reichert
Institute of Databases and Information
   Systems
University of Ulm
Ulm
Germany

Roy Oberhauser
Faculty of Electronics and Computer
   Science
Aalen University
Aalen
Germany

ISSN 1868-4394   ISSN 1868-4408   (electronic)
Intelligent Systems Reference Library
ISBN 978-3-319-84840-2   ISBN 978-3-319-52181-7   (eBook)
DOI 10.1007/978-3-319-52181-7

This Springer imprint is published by Springer Nature
The registered company is Springer International Publishing AG
The registered company address is: Gewerbestrasse 11, 6330 Cham, Switzerland

Contents

About the Editors

Gregor Grambow holds a Ph.D. degree from Ulm University where he worked at the Databases and Information Systems Institute with Manfred Reichert. Before that, he worked at the Computer Science Department together with Roy Oberhauser. Gregor also has a Diploma degree from Aalen University of Applied Sciences and a M.Sc. degree from Karlsruhe University of Applied Sciences. Besides, he has practical experience with software projects from working in different bigger companies in the Karlsruhe high technology region. His research interests include dynamic processes, software engineering, and automatic support for humans. Currently, he is working as BPM consultant at AristaFlow GmbH.

Roy Oberhauser is Professor of Computer Science at Aalen University of Applied Sciences in Germany, teaching in the areas of software and systems engineering. Prior to this, he worked at various companies in the software industry both in the Silicon Valley and in Germany doing research and development. His current applied research interest lies in the area of software engineering for leveraging technologies and techniques to innovate, automate, support, and improve the production and quality of software for society.

Manfred Reichert is Professor at the University of Ulm, Germany and Director of the Databases and Information Systems Institute. His major research interests include business process management (e.g., adaptive processes, data-centric processes, knowledge-intensive processes, and mobile processes), service-oriented computing (e.g., service composition), and process-aware information systems in advanced application domains (e.g., automotive engineering). Manfred pioneered the work on the ADEPT process management technology and is co-founder of the AristaFlow GmbH. He has been participating in numerous research projects in the BPM area contributing more than 250 scientific papers on BPM-related topics. His book entitled "Enabling Flexibility in Process-Aware Information Systems", which he co-authored with Barbara Weber, was recently published by Springer. Manfred has been PC Co-Chair of the BPM'08, CoopIS'11, and EDOC'13 conferences and General Chair of the BPM'09 and EDOC'14 conferences.

Chapter 1
On the Fundamentals of Intelligent Process-Aware Information Systems

Gregor Grambow, Roy Oberhauser and Manfred Reichert

Abstract Process-aware information systems (PAIS) are utilized in business and industry world-wide, and processes constitute a valuable knowledge asset for companies. While the enterprise landscape is subjected to a high rate of change, with developments such as globalization demanding new approaches and technologies like cloud computing, big data, mobile applications, or event processing, PAIS have not kept pace with these trends to adequately manage processes. This article gives an overview of the challenges and novel developments of various systems that have focused on addressing this area, specifically approaches that relate to the most recent category of PAIS we call Intelligent Process-Aware Information Systems (IPAIS).

Keywords Intelligent process-aware information systems · IPAIS · Intelligent business process management systems · IBPMS · Smart processes

1.1 Introduction

Competition is fierce for companies in many sectors today, and a significant challenge they face is the increasingly complex environments in which they must operate. Companies collaborate in complex supply chains, and customers and regulators have numerous requirements the companies must conform to. They may need to handle large amounts of data, coordinate collaborations with a number of partner compa-

G. Grambow (✉) · M. Reichert
Institute of Databases and Information Systems, Ulm University, Ulm, Germany
e-mail: gregor.grambow@uni-ulm.de
URL: http://www.uni-ulm.de/dbis

M. Reichert
e-mail: manfred.reichert@uni-ulm.de

R. Oberhauser
Computer Science Department, Aalen University, Aalen, Germany
e-mail: roy.oberhauser@hs-aalen.de

© Springer International Publishing AG 2017
G. Grambow et al. (eds.), *Advances in Intelligent Process-Aware
Information Systems*, Intelligent Systems Reference Library 123,
DOI 10.1007/978-3-319-52181-7_1

1

nies, incorporate contextual data, manage human-centric and knowledge-intensive processes, integrate or incorporate social media aspects, etc.

This in turn affects company processes, often with a corresponding increase in complexity. To be able to adequately support these complex processes, the information technology (IT) solutions applied in these companies must keep pace with these developments, lest they hinder and confine the companies' capabilities and competitiveness. Especially Process-Aware Information Systems (PAIS) must continually adjust and evolve their processes to be able to handle these new influences and to ensure their capability to model, enact, and support the real-world processes.

These PAIS developments have not gone unnoticed, and leading market research institutes predict a quickly growing need and market for these new process applications. For example, Gartner Research calls these intelligent business process management suites. To account for these developments, they have exchanged their magic quadrant for business process management (BPM) suites [1] with a magic quadrant for intelligent business process suites (IBPM) from 2012 on. In their 2012 version [2] of the magic quadrant they explicitly mention features of IBPM suites that distinguish them from BPM suites. Examples of these are social capabilities, complex event processing (CEP), various types of analyses, or support for business rules. In both the 2012 and the 2014 [3] versions they also name key capabilities of IBPM system (IBPMS) suites, like a graphical modeling environment, content handling, human interactions, business rules, or process repositories.

Like Gartner, Forrester Research also published a study on this topic. They call the new generation of tools smart process applications [4]. They highlight the special properties intelligent processes need nowadays: dynamicity, collaboration, human-centricity, structured and unstructured processes, reporting, transparency, or reporting. They also set up a Forrester Wave for Smart Business Applications [5] and stress the importance of the human as well as capabilities for data and context management. In addition to that, the Forrester Wave for BPM suites [6] states that human-centric, document-centric, and integration-centric BPMs merge into one these days. The key capabilities of these systems involve social processes, cross-organizational processes, collaboration, guidance, and cloud-based services and processes.

Keep in mind that various solutions, tools, and concepts are already in the market and represent the state-of-the-art in applied BPM. The focus of this book goes beyond the current BPM horizon, and considers potential approaches and solutions that could provide insights into and support the future of intelligent processes. Therefore, this book describes fundamentals related to Intelligent Process-Aware Information Systems (IPAIS), their lifecycle, and various advanced approaches and solutions for supporting automated intelligent processes in the complex environments of today and tomorrow.

1.1.1 Evolution of Process Management Systems

In the field of process management systems there has been a substantial amount of scientific work during the last decades. Many terms have been used, including Workflow Management System (WfMS), Business Process Management System (BPMS), or PAIS. In this section, we will give a brief overview about the evolution of these systems.

In the 1990s the term workflow management was popular. Process thinking was already established in companies and WfMS were the systems that brought process-orientation to business information systems. In [7], the evolution of business information systems is discussed: in the 1960s, business information systems were mostly stand-alone systems that were implemented on-top of an operating system. Each of these systems had their own custom-implemented data and user handling. In the 1970s, data management was optimized by introducing dedicated systems for it, the database management systems (DBMS). The same happened in the 1980s to the user interfaces with dedicated user interface management systems (UIMS). Finally, in the 1990s, WfMS enabled the dedicated management of parts of company processes as workflows that were automatically governed by these systems. According to the Workflow Management Coalition (WfMC), a workflow is defined as follows [8]:

> The automation of a business process, in whole or part, during which documents, information or tasks are passed from one participant to another for action, according to a set of procedural rules.

Accordingly, a WfMS is defined as [8]:

> A system that defines, creates, and manages the execution of workflows through the use of software, running on one or more workflow engines, which is able to interpret the process definition, interact with workflow participants and, where required, invoke the use of IT tools and applications.

WfMS were a first step towards implementing, automating, and supporting processes. However, their primary focus was the structural governance of the sequencing of different activities, with the core functionalities being the implementation and enactment of the workflows that were part of a process. They also provided modeling facilities, but this considered only a part of an overall process design. The same applies for diagnosis and analysis of processes. Monitoring capabilities, when included, had only limited functionality. Thus, WfMS did not manage to implement the entire BPM lifecycle as illustrated in Fig. 1.1. This lifecycle consists of the design, the implementation, the enactment, and the diagnosis or analysis of the process.

In the 2000s, two other terms came into the focus: BPMS and PAIS. They are both mostly used for systems that are an extension of WfMS. [9] defines BPM as follows:

> Supporting business processes using methods, techniques, and software to design, enact, control, and analyze operational processes involving humans, organizations, applications, documents and other sources of information.

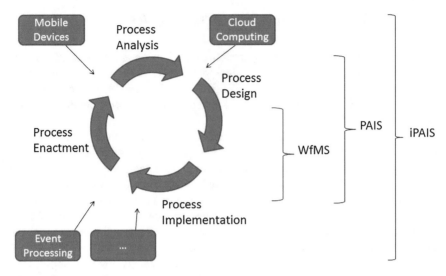

Fig. 1.1 Evolution of process management systems

[10] provides a definition for PAIS that is in line with the preceding definition. A PAIS is

> a software system that manages and executes operational processes involving people, applications, and/or information sources on the basis of process models.

The focus of PAIS was broader than that of WfMS, extending those functionalities with process analysis [9] and enabling better integration and user support. For example, they mostly incorporated more advanced facilities for processing modeling and design, including different features such as organizational models for the processes.

However, in more recent times, new challenges emerged for such systems due to various developments, such as increased organizational and business communication and interaction, new technologies, or the amount of data integration and processing. In Fig. 1.1, a selection of these challenges is presented. One example are the various mobile devices like smartphones or tablets. Many users possess such devices and have an affinity to the mobility and specific user interfaces they have. Thus, a partly mobile implementation of processes comes into play. Another topic gaining momentum is event processing. The myriad of different events important for a company and its processes are recorded and used by information systems (IS). A third example is the trend towards cloud computing: the dynamic provisioning of services using this technology becomes more relevant for integration with processes.

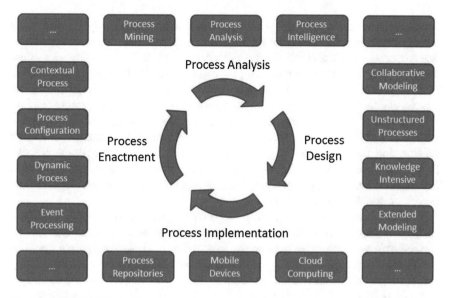

Fig. 1.2 The IPAIS lifecycle

1.2   The IPAIS Lifecycle

In alignment with the BPM lifecycle mentioned in various publications, this section introduces the IPAIS lifecycle as an extension of the BPM lifecycle. An IPAIS extends the capabilities of a PAIS and at least partially incorporates various technologies that can play an important role in the future technical environment. Figure 1.2 shows the different phases of the IPAIS lifecycle with a selection of such technologies. Each phase will be described in a separate subsection and provide examples for approaches and technologies that IPAIS can provide.

1.2.1   Intelligent Process Design

Supply chains, companies, products, and projects grow, while the technologies involved become increasingly complex. This results in process modeling and design becoming complex, while companies strive to cover and automate significantly more real-world processes. Many processes are dynamic at runtime or even completely unstructured, and can no longer be modeled as linearly and statically as, for example, production processes once were. For such processes, the prescriptive way of modeling provided by WfMS or PAIS is not suitable. However, there exist approaches to support users in modeling these processes. One example is DECLARE [11], which is targeted at dynamic, unstructured processes. The approach is based on constraints, and thus the models do not prescribe the exact flow of the process. They rather describe situations in which certain activities can or cannot be executed. At runtime, each activity that does not violate the constraints can be executed. To model processes that are somewhat more structured but still must adapt to different situations, process configuration approaches like C-EPC [12] can be suitable. They enable integration of configurable elements into the process models so that specific fragments of the process can be selected according to parameters that represent the current situation.

Although they are an important asset of organizations, processes were often viewed in isolation. Their connection to and impact on goals, strategies, and other organization assets was often neglected. Today, advanced modeling approaches like [13] based on semantic web technologies exist, which enable the modeled processes to incorporate various types of additional data into the models, better integrating the processes into their organization.

Another feature IPAIS can provide is better support for the complicated modeling of contemporary processes. Examples are advanced correctness checks or a correctness-by-construction approach as provided by [14]. Such approaches prevent the creation of erroneous models. However, process modeling mostly involves significant knowledge and is time consuming. To support reuse of such knowledge, approaches like ProCycle [15] assist the modeler with integrated process lifecycle support. As processes today often involve multiple parties, these are also involved in creating the complex models. Thus, collaborative modeling can substantially reduce modeling efforts. Approaches for dealing with this issue exist, for an overview see [16].

1.2.2 Intelligent Process Implementation

Process implementation refers to the technical realization of the modeled processes to enable their automated enactment. In this areas, various new challenges emerged. With the increasing use of IT systems in organizations, the number of processes enacted by PAIS are also steadily increasing. To manage large numbers of processes effectively, process repositories have been developed. For example, Apromore [17] is a complete process model repository with functionalities for analysis, management, and usage of large sets of process models. However, more specific approaches dealing with specific aspects of such repositories exist. Examples include the querying and retrieval of the models. To support this, [18] provides an approach for visually querying graph-based process models. However, the process models contained in a repository have to be maintained and adapted according to real-world events from time to time, whereby redundancies might be introduced and the maintainability of the models endangered. To counteract this, [19] provides a catalog of 'process model smells' and behavior-preserving refactorings.

As companies collaborate globally, availability of the processes is also a key implementation issue. In this area, cloud computing is on the rise and there are IPAIS approaches combining BPM and cloud techniques. In [20], a novel approach is proposed that supports the distribution of end-user activities via the cloud. Furthermore, the transfer of sensitive data is also addressed. The approach shown in [21], in turn, combines BPM, service-oriented architecture (SOA), and cloud technology to improve the availability of services, in this case for the educational sector.

Another important trend in many domains is the integration of mobile devices in processes. To support this with technology, portions of these processes may be implemented and enacted on these devices. For this purpose IPAIS approaches have been developed that enable partly mobile implementations, such as DEMAC [22], ROME4U [23], or MARPLE [24]. They include features like the transfer of parts of a process to a mobile device even at run-time.

1.2.3 Intelligent Process Enactment

The PAIS area probably affected the most by current challenges and technologies is process enactment. Real-world processes involving humans are mostly dynamic and IPAIS offer ways to align such processes to changing situations. One option for this is the configuration of enactable processes as done in Provop [25]. Some approaches even allow for automatic, correct, and context-sensitive configuration without human involvement [26]. A related area of research seeks to avoid overburdening users with large process models during execution by not configuring the process model but only the part of it that the user is shown. An example is ProView [27, 28], which enables the creation of user-based process views in which changes can be applied that are propagated back to the base process instance.

For even more dynamic scenarios, adaptive process approaches can be utilized, which have been utilized within the scientific community for some time [29–31]. Beyond scientific processes, a prominent approach for run-time adaptation is the ADEPT project [14] that includes correctness guarantees. On this basis, other approaches have been developed that even automate the run-time adaptations of the processes for exception handling [32] or for dynamically inserting software development quality assurance measures [33].

Another aspect gaining importance these days is the context of processes. This incorporates various influences like humans, artifacts, or events occurring while processes are enacted. For most approaches, events and context correspond. However, despite the availability of context management frameworks like [34] or [35], few approaches combine these directly with process enactment. One is presented in [36] that combines a CEP system and a PAIS. Another approach in this area deals with context building and sharing in the context of mobile web services [37]. Artifacts that are tied to the processes are, however, treated explicitly by artifact-centric PAIS approaches. Examples here include [38] and [39]. Both manage the lifecycle of important artifacts and relate it to the processes.

1.2.4   Intelligent Process Diagnosis

PAIS has extended WfMS primarily in the area of process diagnosis and analysis. However, these functionalities were mostly limited to monitoring and simple analyses. In the early 2000s, process mining [40] focused on the discovery of processes utilizing event logs. In particular, the control flow of these processes was the focus. However, mining techniques yield the potential for many types of analyses. One example is social network mining [41] that focuses on the connections of the different participating actors. Similar to this, organizational mining [42] targets the mining of the organizational structures behind the actors in a process.

IPAIS, in turn, goes a step further with more advanced capabilities. In [43] mining is applied for conformance checking of processes. In particular, security audits in the financial sector are applied with mining techniques. Another advanced application [44] enables the prediction of the completion time of future processes based on past instances. As processes are subject to ongoing changes, [45] provides an approach for mining the different executed variants of a process model. Furthermore, the most suitable changes to the process model can be derived. Mining techniques can still be further extended and improved. [46] proposes an approach for comprehensive process diagnostics based on process mining. This includes the control flow perspective, the performance perspective, as well as the organizational perspective. Another interesting extension is semantic process mining [47]. This technique uses semantic annotations in the processes to enable more advanced and comprehensive analyses.

1.2.5   IPAIS Approaches Discussed in This Book

This section provides a brief overview of the specific IPAIS approaches covered in this book. The focus of these approaches covers different IPAIS areas like the presentation of entire IPAIS tools, the delivery of process-relevant information to the users during enactment, and different kinds of analyses conducted on processes.

1.2.5.1   Adaptive Process Management

In today's world, processes are enacted in many domains. Most of them are not as predictable as standardized industrial production processes. They rather map a real world process that is subjected to many often unforeseen influences. Systems that operate in this context are often called cyber-physical systems, as they apply computational elements to control real-world physical entities. Such systems struggle with a vast set of exceptional situations. To achieve better support for the latter, Chap. 2 proposes a system that is capable of automated process adaptations. Furthermore, it can detect exceptional situations and failures and thus apply matching adaptations.

To achieve this, a declarative task specification and AI management techniques are applied to avoid the need for specifying exception handlers at build-time.

1.2.5.2   Case Management Processes

Case management is an approach often used in the context of PAIS and IPAIS. It suits human-centric and dynamic processes well, as it does not focus on the imperative control-flow of a process, but rather focuses on the activities to complete for a specific case and the actors that required to perform these. The approach presented in Chap. 3 is positioned in this area. Its contribution is twofold. First, case management processes are concertized by a formalism that is based on hierarchical state machines and state charts. The latter are also extended to better suit the case management topic. Based on this, an architecture of a case management support system is discussed. This system shall facilitate case management by context-driven recommendations for activity planning.

1.2.5.3   Autonomically-Capable Processes

PAIS are becoming more important as processes gain in their standing as valuable assets of companies. However, both are coupled with substantial costs for the companies due to high efforts for design, evaluation, optimization, or adaptation of the processes. In recent years, there has been a trend towards autonomic computing, making applications more independent and automated and reducing human intervention to a minimum. Chap. 4 presents the vision and challenges of autonomically-capable processes, applying the properties of autonomic computing to IPAIS, including self-configuration, self-optimization, and other self-X properties. Additionally, the state-of-the-art in this area is presented and reviewed. Finally, an example of a concrete system example that realizes many of the discussed properties is presented.

1.2.5.4   Process-Oriented Information Logistics

In today's companies, knowledge-workers are confronted with a steadily rising amount of data, information, and artifacts. These knowledge-workers are taking part in various processes considered valuable to the companies. However, the workers are struggling with different impediments concerning their work. The information and data overload makes it difficult to effectively and efficiently manage and use the different artifacts. This problem is amplified by the fact that the process and the relevant information are mostly managed in different information systems. The approach presented in Chap. 5 addresses this problem. It automates delivery of information to the user, selecting information in a process-oriented and context-aware manner to fit the situation of the knowledge-worker. In addition to presenting the concept, this chapter also discusses concrete use cases and a proof-of-concept prototype.

1.2.5.5   Process Recommendations

Process-related data plays an important role in various enterprises. Such data must be retrieved periodically or in real-time from various data sources and systems. Further, it must be aggregated to KPIs (Key Performance Indicators). That way decision makers are enabled to continuously monitor the business processes in order to maintain high process, product, and service quality. In general, a retrospective analysis of process data is conducted to highlight violated KPIs. However, this cannot ensure prevention of future KPI violations. To counteract this problem, PAIS should also enable prospective analysis of processes to identify processes not performing as intended. Therefore, Chap. 6 presents both a methodology and architecture for predictive process analysis. This is enabled by comparing running process instances with historic data via machine learning algorithms.

1.2.5.6   Reasoning over Process Models

Visual modeling languages like BPMN (Business Process Modeling Notation) have established themselves as a standard for process modeling. However, besides their many advantages, their semi formal nature together with the use of natural language for parts of the processes can introduce problems. Natural languages leave room for interpretation and ambiguities, like homonyms or synonyms. In particular, in the context of collaborative modeling or process model sharing, this can lead to misunderstandings that might even endanger the success of the process. Furthermore, process modeling languages lack machine readable semantics describing the content of the models. Thus, tool-aided support for process modeling and execution is inevitably limited. The contribution of Chap. 7 is situated in the area of semantic processes and addresses these problems. In particular, semantic web technology such as ontologies and reasoners are applied to achieve an extension of process descriptions containing machine readable semantics. In addition to the abstract concept, the chapter presents a proof-of-concept tool for annotating and converting process descriptions that also allows for querying the ontology-based process description.

1.2.5.7   Improving Process Portability

Today's enterprise landscape is subjected to ongoing change including process technologies. Recent trends include various implementation aspects of processes, such as cloud- or service-based PAIS. Such constantly changing technologies, however, promote frequent changes in the applied tooling of companies. In light of this, it becomes important that process specification be done in a way that facilitates their portability to another PAIS without necessitating a complete reimplementation. Adherence to standards like BPMN is one option to support this. However, BPM vendors interpret and implement such standards differently. Thus, portability is hampered and in many cases processes cannot be transferred from one PAIS to another. To tackle this problem, Chap. 8 provides an approach based on metrics for process portability. The contribution of the chapter is two-fold: first, the approach can be used to aid the decision whether to port or to rewrite a process-based application. Second, it can be integrated into the development process to support the creation of more portable processes.

1.2.5.8   Business Process Intelligence

Business processes are an important asset of today enterprises. To maximize the value derived from this asset, one emerging trend in the business intelligence world is BPI (Business Process Intelligence). In particular, BPI is a vast area consisting of various approaches. This comprises methods and best practices from real-time process analysis as well as concrete software tools applying them. Chapter 9 analyzes contemporary software tools enabling BPI. Therefore, the chapter presents an analysis of the features as well as the strengths and weaknesses of each tool. Finally, the chapter discusses two application strategies for BPI software tools in modern enterprises.

1.3   Outlook

PAIS will evolve to include IPAIS technology that is dynamically flexible and can adapt to changing contexts and issues while reducing human intervention and costly administrative and management overhead.

We thus foresee IPAIS as gaining in significance and providing key automation capabilities for addressing the efficiency and effectiveness of business and industry processes improving our society. Before that vision can happen, current PAIS technology needs to fulfill important autonomic capabilities and enhance the process lifecycle, and integrating various technologies and approaches as described in this book can bring us incrementally closer to turning this vision into reality.

References

1. Sinur, J., Hill, J.B.: Magic quadrant for business process management suites. Gartner RAS Core Research Note, pp. 1–24 (2010)
2. Sinur, J., Schulte, W., Hill, J., Jones, T.: Magic quadrant for intelligent business process management suites. Gartner, Stamford, CT. https://www.gartner.com/doc/2179715 (2012). Accessed Dec 2014
3. Jones, T., Schulte, W., Cantara, M.: Magic quadrant for intelligent business process management suites. Gartner, Stamford, CT. https://www.gartner.com/doc/2684315/ (2014). Accessed Dec 2014
4. Bartels, A., Moore, C., Mines, C., Warner, J., Hamerman, P., Le Clair, C., Band, W., Yamnitzki, M., J, C.: Smart process applications fill a big business gap. Gartner, Stamford, CT. https://www.forrester.com/Smart+Process+Applications+Fill+A+Big+Business+Gap/fulltext/-/E-RES77442 (2014). Accessed Dec 2014

5. Bartels, A., Moore, C., Mines, C., Le Clair, C., Richardson, C., Miers, D., J, C.: The forrester wave: smart process applications, Q2 2013. Gartner, Stamford, CT. https://www.forrester.com/The+Forrester+Wave+Smart+Process+Applications+Q2+2013/fulltext/-/E-RES82923 (2013). Accessed Dec 2014

6. Richardson, C., Miers, D., A, C., Keenan, J.: The forrester wave: BPM suites, Q1 2013. Gartner, Stamford, CT. https://www.forrester.com/The+Forrester+Wave+BPM+Suites+Q1+2013/fulltext/-/E-RES88581 (2013). Accessed Dec 2014

7. van der Aalst, W.M.: The application of Petri nets to workflow management. J. Circ. Syst. Comput. **8**(01), 21–66 (1998)

8. WFMC: Terminology and glossary (WFMC-TC-1011). Technical report. Workflow Management Coalition, Brussels (1996)

9. Van Der Aalst, W.M., Ter Hofstede, A.H., Weske, M.: Business process management: a survey. In: Business Process Management, pp. 1–12. Springer (2003)

10. Dumas, M., Van der Aalst, W.M., Ter Hofstede, A.H.: Process-aware Information Systems: Bridging People and Software Through Process Technology. Wiley (2005)

11. Pesic, M., Schonenberg, H., van der Aalst, W.M.: Declare: Full support for loosely-structured processes. In: 11th IEEE International Enterprise Distributed Object Computing Conference, 2007. EDOC 2007, p. 287. IEEE (2007)

12. Rosemann, M., van der Aalst, W.M.P.: A configurable reference modelling language. Inf. Syst. **32**(1), 1–23 (2005)

13. Markovic, I.: Semantic Business Process Modeling. KIT Scientific Publishing (2010)

14. Dadam, P., Reichert, M.: The ADEPT project: a decade of research and development for robust and flexible process support—challenges and achievements. Comput. Sci. Res. Dev.**23**(2), 81–97 (2009)

15. Weber, B., Reichert, M., Wild, W., Rinderle-Ma, S.: Providing integrated life cycle support in process-aware information systems. Int. J. Coop. Inf. Syst. (IJCIS) **18**(1), 115–165 (2009)

16. Renger, M., Kolfschoten, G.L., de Vreede, G.J.: Challenges in collaborative modeling: a literature review. In: Advances in Enterprise Engineering I, pp. 61–77. Springer (2008)

17. La Rosa, M., Reijers, H.A., Van Der Aalst, W.M., Dijkman, R.M., Mendling, J., Dumas, M., GarcíA-BañUelos, L.: Apromore: an advanced process model repository. Expert Syst. Appl. **38**(6), 7029–7040 (2011)

18. Awad, A., Sakr, S.: Querying graph-based repositories of business process models. In: Database Systems for Advanced Applications, pp. 33–44. Springer (2010)

19. Weber, B., Reichert, M., Mendling, J., Reijers, H.A.: Refactoring large process model repositories. Comput. Ind. **62**(5), 467–486 (2011)

20. Han, Y.B., Sun, J.Y., Wang, G.L., Li, H.F.: A cloud-based BPM architecture with user-end distribution of non-compute-intensive activities and sensitive data. J. Comput. Sci. Technol. **25**(6), 1157–1167 (2010)

21. Mircea, M.: SOA, BPM and cloud computing: connected for innovation in higher education. In: International Conference on Education and Management Technology (ICEMT), 2010, pp. 456–460. IEEE (2010)

22. Zaplata, S., Hamann, K., Kottke, K., Lamersdorf, W.: Flexible execution of distributed business processes based on process instance migration. J. Syst. Integr. **1**(3), 3–16 (2010)

23. Russo, A., Mecella, M., de Leoni, M.: ROME4EU-A service-oriented process-aware information system for mobile devices. Softw.-Pract. Experience **42**(10), 1275–1314 (2012)

24. Pryss, R., Musiol, S., Reichert, M.: Collaboration support through mobile processes and entailment constraints. In: 9th IEEE International Conference on Collaborative Computing: Networking, Applications and Worksharing (CollaborateCom'13). IEEE Computer Society Press, Oct 2013

25. Hallerbach, A., Bauer, T., Reichert, M.: Capturing variability in business process models: The provop approach. J. Softw. Maint. Evol.: Res. Pract. **22**(6–7), 519–546 (2010)

26. Grambow, G., Mundbrod, N., Kolb, J., Reichert, M.: Towards simple and robust automation of sustainable supply chain communication. In: CoopIS. Number 8842 in LNCS, pp. 644–647. Springer (2014)

27. Kolb, J., Kammerer, K., Reichert, M.: Updatable process views for user-centered adaption of large process models. In: 10th International Conference on Service Oriented Computing (ICSOC'12). Number 7636 in LNCS, pp. 484–498. Springer, Oct 2012
28. Kolb, J., Reichert, M.: Data flow abstractions and adaptations through updatable process views. In: 28th Symposium on Applied Computing (SAC'13), 10th Enterprise Engineering Track (EE'13), pp. 1447–1453. ACM Press, Mar 2013
29. Sadiq, S.W., Marjanovic, O., Orlowska, M.E.: Managing change and time in dynamic workflow processes. Int. J. Coop Inf. Syst. **9**(01–02), 93–116 (2000)
30. Weske, M.: Formal foundation and conceptual design of dynamic adaptations in a workflow management system. In: Proceedings of the 34th Annual Hawaii International Conference on System Sciences, 2001, p. 10. IEEE (2001)
31. Bandinelli, S., Fuggetta, A., Ghezzi, C., Lavazza, L.: Spade: An environment for software process analysis, design, and enactment. Softw. Process Modell. Technol. 223–247 (1994)
32. Müller, R., Greiner, U., Rahm, E.: Agentwork: a workflow system supporting rule-based workflow adaptation. Data Knowl. Eng. **51**(2), 223–256 (2004)
33. Grambow, G., Oberhauser, R., Reichert, M.: Contextual injection of quality measures into software engineering processes. Int. J. Adv. Softw. **4**(1&2), 76–99 (2011)
34. Fahy, P., Clarke, S.: CASS—a middleware for mobile context-aware applications. In: Workshop on Context Awareness, MobiSys, Citeseer (2004)
35. Gu, T., Pung, H.K., Zhang, D.Q.: A middleware for building context-aware mobile services. In: 2004 IEEE 59th Vehicular Technology Conference, 2004. VTC 2004-Spring, vol. 5, pp. 2656–2660. IEEE (2004)
36. Von Ammon, R., Emmersberger, C., Springer, F., Wolff, C.: Event-driven business process management and its practical application taking the example of DHL. In: 1st International Workshop on Complex Event Processing for the Future Internet (2008)
37. Dorn, C., Dustdar, S.: Sharing hierarchical context for mobile web services. Distrib. Parallel Databases **21**(1), 85–111 (2007)
38. Bhattacharya, K., Hull, R., Su, J., et al.: A data-centric design methodology for business processes. In: Handbook of Research on Business Process Modeling, pp. 503–531 (2009)
39. Künzle, V., Reichert, M.: PHILharmonicFlows: towards a framework for object-aware process management. J. Softw. Maint. Evol.: Res. Pract. **23**(4), 205–244 (2011)
40. Van der Aalst, W.M., Weijters, A.: Process mining: a research agenda. Comput. Ind. **53**(3), 231–244 (2004)
41. Van der Aalst, W.M., Song, M.: Mining social networks: uncovering interaction patterns in business processes. In: Business Process Management, pp. 244–260. Springer (2004)
42. Song, M., van der Aalst, W.M.: Towards comprehensive support for organizational mining. Decis. Support Syst.**46**(1), 300–317. (2008)
43. Accorsi, R., Stocker, T.: On the exploitation of process mining for security audits: the conformance checking case. In: Proceedings of the 27th Annual ACM Symposium on Applied Computing, pp. 1709–1716. ACM (2012)
44. van der Aalst, W.M., Schonenberg, M.H., Song, M.: Time prediction based on process mining. Inf. Syst. **36**(2), 450–475 (2011)
45. Li, C., Reichert, M., Wombacher, A.: Mining business process variants: challenges, scenarios, algorithms. Data Know. Eng. **70**(5), 409–434 (2011)
46. Bozkaya, M., Gabriels, J., Werf, J.: Process diagnostics: a method based on process mining. In: International Conference on Information, Process, and Knowledge Management, 2009. eKNOW'09, pp. 22–27. IEEE (2009)
47. De Medeiros, A.A., Pedrinaci, C., van der Aalst, W.M., Domingue, J., Song, M., Rozinat, A., Norton, B., Cabral, L.: An outlook on semantic business process mining and monitoring. In: On the Move to Meaningful Internet Systems 2007: OTM 2007 Workshops, pp. 1244–1255. Springer (2007)

Chapter 2
Adaptive Process Management in Cyber-Physical Domains

Andrea Marrella and Massimo Mecella

Abstract The increasing application of process-oriented approaches in new challenging cyber-physical domains beyond business computing (e.g., personalized healthcare, emergency management, factories of the future, home automation, etc.) has led to reconsider the level of flexibility and support required to manage complex processes in such domains. A cyber-physical domain is characterized by the presence of a cyber-physical system coordinating heterogeneous ICT components (PCs, smartphones, sensors, actuators) and involving real world entities (humans, machines, agents, robots, etc.) that perform complex tasks in the "physical" real world to achieve a common goal. The physical world, however, is not entirely predictable, and processes enacted in cyber-physical domains must be robust to unexpected conditions and adaptable to unanticipated exceptions. This demands a more flexible approach in process design and enactment, recognizing that in real-world environments it is not adequate to assume that all possible recovery activities can be predefined for dealing with the exceptions that can ensue. In this chapter, we tackle the above issue and we propose a general approach, a concrete framework and a process management system implementation, called SmartPM, for automatically adapting processes enacted in cyber-physical domains in case of unanticipated exceptions and exogenous events. The adaptation mechanism provided by SmartPM is based on declarative task specifications, execution monitoring for detecting failures and context changes at run-time, and automated planning techniques to self-repair the running process, without requiring to predefine any specific adaptation policy or exception handler at design-time.

Keywords Process adaptation · Process management system · Cyber-physical system · Emergency management · Knowledge representation · Automated planning

A. Marrella (✉) · M. Mecella
Sapienza Università di Roma, Rome, Italy
e-mail: marrella@diag.uniroma1.it

M. Mecella
e-mail: mecella@diag.uniroma1.it

A. Marrella · M. Mecella
Dipartimento di Ingegneria Informatica Automatica e Gestionale, Rome, Italy

© Springer International Publishing AG 2017
G. Grambow et al. (eds.), *Advances in Intelligent Process-Aware Information Systems*, Intelligent Systems Reference Library 123,
DOI 10.1007/978-3-319-52181-7_2

2.1   Introduction

As Information and Communication Technologies (ICTs) are being increasingly integrated and embedded into our everyday environment, the design of embedded ICT from components (PCs, smartphones, sensors, actuators, etc.) to *cyber-physical systems* is becoming a reality. A cyber-physical system (CPS) is a system of *interconnected* and *collaborating* computational elements controlling physical components that provide real world entities (e.g., humans, machines, agents, robots, etc.) with a wide range of innovative applications and services [1]. CPSs are designed to support and facilitate collaboration among people and software services on complex tasks. On the other side, the Business Process Management (BPM) discipline has gained an increasing importance in describing complex correlations between distributed systems and offers a powerful representation of collaborative activities [2]. In the field of online trading and manufacturing, for example, modelling and execution languages for business processes, such as BPMN [3] and BPEL [4], have proven to be well suited to formalize high-level sequences of activities involving web service invocations and human interaction.

Nowadays, the current maturity of process management systems (PMSs) and methodologies has led to the application of process-oriented approaches in new challenging *cyber-physical domains* beyond business computing [5, 6], such as personalized healthcare [7–9], emergency management [10, 11], factories of the future [12] and home automation [13]. Such domains are characterized by the presence of a CPS coordinating heterogeneous ICT components with a large variety of architectures, sensors, computing and communication capabilities, and involving real world entities that perform complex tasks in the "physical" real world to achieve a common goal. In this context, a PMS is used to manage the life cycle of the collaborative processes that coordinate the services offered by the CPS to the real world entities. To guarantee a better control over the interaction that PMS has with the real world, it continuously collects contextual information from the specific cyber-physical domain it is employed in.

The long-term objective of CPSs is to create a strong link between the physical world and the cyber world to support their users with performing their tasks [14]. The physical world, however, is not entirely predictable. CPSs do not necessarily and always operate in a controlled environment, and their collaborative processes must be robust to unexpected conditions and adaptable to exceptions and external exogenous events. To this end, we define an *exception* as any deviation from an "ideal" collaborative process that uses the available resources to achieve the task requirements in an optimal way [15].

Exception handling is one of the most important tasks that process designers undertake during business process modelling and execution [16]. Exceptions can be either anticipated or unanticipated. An anticipated exception can be planned at design-time and incorporated into the process model, i.e., a (human) process designer can provide an exception handler which is invoked during run-time to cope with the exception. Conversely, unanticipated exceptions generally refer to situations, unplanned at design-time, that may emerge at run-time and can be detected by monitoring discrepancies and inconsistencies between the real-world processes and their computerized representation. To cope with those exceptions, a PMS is required to allow ad hoc process changes for adapting running process instances in a situation- and context-dependent way.

However, in cyber-physical domains, the number of possible anticipated exceptions is often too large, and traditional manual implementation of exception handlers at design-time is not feasible for the process designer, who has to anticipate all potential problems and ways to overcome them in advance. Furthermore, anticipated exceptions cover only partially relevant situations, as in such scenarios many unanticipated exceptional circumstances may arise during the process execution. While most PMSs of today shy away from dealing with the inherent dynamic nature of cyber-physical domains [12], the management of processes enacted in such domains requires a PMS providing real-time monitoring and adaptation features during process execution. This requires the formalization of explicit mechanisms to model world changes and responding to anomalous situations, exceptions, exogenous events in an *automated* way, in order to achieve the overall objectives of the processes still preserving their structure without (or by minimising) any human intervention.

In this chapter, we tackle the above challenge by presenting a general approach, a concrete framework and a PMS implementation, called SmartPM (Smart Process Management) for automatically adapting processes enacted in cyber-physical domains in case of unanticipated exceptions and exogenous events. SmartPM is based on declarative task specifications, process execution monitoring for detecting failures and context changes at run-time, and automated exception handling and resolution strategies on the basis of well-established Artificial Intelligence (AI) techniques. Even more importantly, the adaptation mechanisms provided by SmartPM allow deviations at run-time from the execution path prescribed by the original process without altering its process model, a feature that makes SmartPM particularly suitable for managing processes in cyber-physical domains.

The rest of the chapter is organized as follows. In Sect. 2.2 we describe the state-of-the-art approaches to process adaptation, by investigating existing techniques to deal with anticipated and unanticipated exceptions in BPM. In Sect. 2.3, we first present a concrete running example of a process enacted in a cyber-physical environment; then, we derive a list of characterizing features that a PMS managing processes in cyber-physical domains should provide. To meet the identified features, in Sect. 2.4 we introduce the general approach to handle with unanticipated exceptions and exogenous events as defined in the SmartPM framework, and we present the architecture of the implemented SmartPM system. Then, in Sect. 2.5 we provide a critical discussion about the general applicability of the SmartPM approach and we trace the future challenges related to the management of processes in cyber-physical domains. Finally, Sect. 2.6 concludes the chapter.

2.2   Related Work

Over the last years, there was a trend in providing PMSs with a growing support for adapting business processes to deal with exceptions, changing environments and evolving needs [16, 17]. If not detected and handled effectively, exceptions can result in severe impacts on the cost and schedule performance of PMSs [18].

Process adaptation techniques rely on the assumption that exceptions and deviations are detectable [19]. When detection capabilities are provided by the PMSs, mainly in the case of anticipated exceptions, the modeling and execution environment enables process designers to define events, triggers and conditions (e.g., timers, error messages, pre- and post-execution constraints, etc.) whose run-time occurrence or violation is recognized as an exception. When the exceptions and deviations are unanticipated or caused by external factors not under the control of the PMS, users (or external systems) are often allowed to explicitly notify the PMS about the detected exception or deviation.

In this section, we describe the state-of-the-art approaches to process adaptation considering to what extent users are involved in the process of defining exception conditions and handling policies (as summarized in Fig. 2.1), which directly influences the degree of automation provided in the exception resolution and process adaptation stages. Specifically, we first outline traditional exception handling techniques used to deal with anticipated exceptions (Sect. 2.2.1). Then, we review the existing approaches allowing ad hoc process changes for adapting running process instances in case of unanticipated exceptions (Sect. 2.2.2). Finally, we analyze a number of techniques from the field of AI that have been applied to BPM with the aim of increasing the degree of automated process adaptation at run-time (Sect. 2.2.3).

2.2.1   Exception Handling

Initial research efforts addressing the need for exception handling in PMSs can be traced back to the late nineties and early two thousands [15, 20–25]. Although possible sources of anticipated exceptions are different (as outlined in [21, 22],

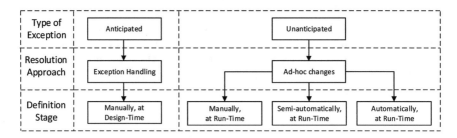

Fig. 2.1 Exception handling and process adaptation approaches

they can be related to activity failures, deadline expirations, resource unavailabilities, constraint violations and external events) and go beyond technical failures, not surprisingly exception handling approaches in PMSs trace and resemble exception handling mechanisms in programming languages. Abstracting from the specific techniques and implementations, a common behavioral pattern can be identified. At design-time, the process designer identifies possible exceptions that may occur, defines exception triggering events and conditions, and specifies exception handlers associated with the predefined process model. Exception handlers can be defined for single activities, for selected process regions (including multiple activities), or for the overall process (as in the case of a `try` block in programming languages). The main process logic is thus clearly separated from the exception handling logic. During process execution, timers, messages, errors, constraint violations and other events might interrupt the process flow: the exception is detected and thrown. The run-time environment checks for the availability of a suitable exception handler, which is then invoked to catch the exception (as in the case of a `catch` block). Typically, the process (or sub-parts of it) is interrupted and the flow of control passes to the exception handler. The handler defines specific activities to be performed to recover from the exception, so that process execution can be possibly resumed.

As extensively discussed in [26], exception handling capabilities provided by academic prototypes and commercial PMSs can be reconducted to the abstract framework introduced before. The different approaches vary in the exception types that can be handled and in the way they support the definition and selection of exception handlers, which can be completely predefined, contextually selected from a repository or instantiated from templates. Typical strategies applied when defining exception handlers for anticipated exceptions have been systematized in the form of *exception handling patterns* [16, 27, 28]. When for a given exception no explicit handling logic is defined or the handler is not able to resolve the issue, a process participant may be notified and involved in the definition of corrective actions.

Several exception detection and handler activation techniques [20, 24, 29] adopt a rule-based approach, typically relying on some form of Event-Condition-Action (ECA) rules. ECA rules have the form "on *event* if *condition* do *action*" and specify to execute the *action* (i.e., the exception handler) automatically when the *event* happens (i.e., when the exception is caught), provided that the specific *condition* holds. ECA rules represent a good way for separating the graphical representation of the process with the "exception handling flow". A similar principle has been applied in YAWL [30], where for each exception that can be anticipated, it is possible to define an exception handling process, named *exlet*, which includes a number of exception handling primitives (for removing, suspending, continuing, completing, failing and restarting a workitem/case) and one or more compensatory processes in the form of *worklets* (i.e., self-contained YAWL specifications executed as a replacement for a workitem or as compensatory processes). Exlets are linked to specifications by defining specific rules (through the Rules Editor graphical tool), in the shape of *Ripple Down Rules* specified as "if condition then conclusion", where the condition defines the exception triggering condition and the conclusion defines the exlet.

2.2.2   Ad Hoc Process Change

Even though the handling of anticipated exceptions is fundamental for every PMS, the latter also needs to be able to deal with unanticipated exceptions. Research efforts dealing with unanticipated exceptions have established the area of *adaptive process management* [16, 31]. While the introduction of exception handling techniques for anticipated exceptions increases process flexibility and adaptation capabilities, a different approach is required for handling unanticipated exceptions and deviations occurring at run-time. The handling of unanticipated exceptions does not assume the availability of predefined exception handlers and relies on the possibility of performing ad hoc changes over process instances at run-time. The need to perform complex behavioral changes over a process instance requires *structural adaptation* of the corresponding process model, which leads to adaptations of the process instance state.

As in the case of exception handling, structural adaptation techniques have been systematized through the identification and definition of adaptation patterns [32, 33]. At a low-level of abstraction, structural model adaptations can be performed by applying change primitives such as adding/removing nodes, routing elements, edges and other process elements. At a higher level of abstraction, change operations provide a set of adaptation patterns to perform model adaptations, such as adding, deleting, moving or replacing activities or process fragments. A single change operation corresponds to the application of multiple change primitives, hiding the complexity of the model editing task. Adaptation patterns are not limited to the control flow perspective and also cover other process perspectives to perform changes, e.g., at the level of the data flow schema or on process resources. In addition, change operations performed for one perspective (e.g., control flow) may affect the other perspectives (e.g., the data flow) as well, resulting in so-called secondary changes. Notice that ad hoc changes must preserve the correctness of the process model and the executability of the process instance [34].

While a good level of support can be provided to ensure correctness and compliance when high-level change operations are performed, the degree of automation in performing these changes is generally limited. In fact, ad hoc changes are often manually performed by experienced users: process execution is suspended and the model and state of the affected instance are adapted by relying on the capabilities of the modeling environment. In an attempt to increase the level of user support, *semi-automated* approaches have been proposed [35]. They aim at storing and exploiting available knowledge about previously performed changed, so that users can retrieve and apply it when adapting a process. Knowledge retrieval and reuse requires establishing a link between performed changes and the application context, including the occurred exception and the process state. Contextual information allows, in turn, identifying similarities between the current exceptional situation and previous cases. The available knowledge on how similar cases were handled in the past is used to assist the users, provide recommendations and suggest possible changes to be applied. Such an approach has been concretely put into practice using case-based reasoning techniques [36, 37].

Strong support for adaptive process management and exception handling is provided by the ADEPT system and its evolutions [38–41]. ADEPTflex offers modeling capabilities to explicitly define pre-specified exceptions, and supports changes of process instances to enable different kinds of ad hoc deviations from the pre-modeled process models in order to deal with run-time exceptions. These features have been extended and improved in ADEPT2, which provides full support for the structural process change patterns defined in [32], and in ProCycle, which combines ADEPT2 with conversational case-based reasoning (CCBR) methodologies. On the basis of the ADEPT technology, the AristaFlow BPM Suite was developed, with the aim of transferring process flexibility and adaptation concepts into an industrial-strength PMS. Similarly, AgentWork [42] relies on ADEPTflex and exploits a temporal ECA rule model to automatically detect logical failures and enable both reactive and predictive process adaptation of control- and data-flow elements. Here, exception handling is limited to single tasks failures, and the possibility exists for conflicting rules to generate incompatible actions, which requires manual intervention and resolution.

If compared with traditional exception handling approaches (cf. Sect. 2.2.1), adaptive PMSs deal with unanticipated exceptions by automatically deriving the try block as the situation in which the PMS does not adequately reflect the real-world process anymore. As a consequence, one or several process instances have to be adapted with ad hoc process changes, and the catch block should include those recovery procedures required for realigning the computerized processes with the real-world ones.

2.2.3   AI-based Process Adaptation

The AI community has been involved with research on process management for several decades, and AI technologies can play an important role in the construction of PMS engines that manage complex processes, while remaining robust, reactive, and adaptive in the face of both environmental and tasking changes [43]. One of the first works dealing with this research challenge is [44]. It discusses at high level how the use of an intelligent assistant based on planning techniques may suggest compensation procedures or the re-execution of activities if some anticipated failure arises during the process execution. In [45] the authors describe how planning can be interleaved with process execution and plan refinement, and investigates plan patching and plan repair as means to enhance flexibility and responsiveness. Similarly, the approach presented in [46] highlights the improvements that a legacy workflow application can gain by incorporating planning techniques into its day-to-day operation.

A goal-based approach for enabling automated process instance change in case of emerging exceptions is shown in [47]. If a task failure occurs at run-time and leads to a process goal violation, a multi-step procedure is activated. It includes the termination of the failed task, the sound suspension of the process, the automatic generation (through the use of a partial-order planner) of a new complete process definition that complies with the process goal and the adequate process resumption. A similar approach is proposed in [48]. The approach is based on learning business activities as planning operators and feeding them to a planner that generates a candidate process model that is capable of achieving some business goals. If an activity fails during the process execution at run-time, an alternative candidate plan is provided with the same business goals. The major issue of [47, 48] lies in the replanning stage used for adapting a faulty process instance, which forces to completely redefine the process specification at run-time when the process goal changes (due to some activity failure), by revolutionizing the work-list of tasks assigned to the process participants (that are often humans).

In the work [49] the authors propose a goal-driven approach for service-based applications to automatically adapt business processes to run-time context changes. Process models include service annotations describing how services contribute to the intended goal, and business policies over domain elements. Contextual properties are modeled as state transition systems capturing possible values and possible evolutions in the case of precondition violations or external events. Process and context evolution are continuously monitored and context changes that prevent goal achievement are managed through an adaptation mechanism based on service composition via automated planning techniques. However, this work requires that the process designer explicitly defines the policies for detecting the exceptions at design-time.

A work dealing with process interference is that of [50]. Process interference is a situation that happens when several concurrent business processes depending on some common data are executed in a highly distributed environment. During the processes execution, it may happen that some of these data are modified causing unanticipated or wrong business outcomes. To overcome this limitation, the work [50] proposes a run-time mechanism that uses (i) *Dependency Scopes* for identifying critical parts of the processes whose correct execution depends on some shared variables; and (ii) *Intervention Processes* for solving the potential inconsistencies generated from the interference, which are automatically synthesised through a domain independent planner based on CSP (*Constraint Satisfaction Problems*) techniques.

2.3 Managing Processes in Cyber-Physical Domains

CPSs are having widespread applicability and proven impact in multiple areas, like aerospace, automotive, traffic management, healthcare, manufacturing, emergency management, entertainment, and consumer appliances [14, 51]. According to [1], any physical environment that contains computing-enabled devices can be considered as a cyber-physical domain. The trend of managing processes in cyber-physical domains has been fueled by two main factors. On the one hand, the recent development of powerful mobile computing devices providing wireless communication capabilities have become useful to support mobile workers to execute tasks in such dynamic settings [52]. On the other hand, the increased availability of sensors disseminated in the world has lead to the possibility to monitor in detail the evolution of several real-world objects of interest. The knowledge extracted from such objects allows to depict the contingencies and the context in which processes are carried out, by consenting a fine-grained monitoring, mining, and decision support for them.

However, if compared with traditional business domains, additional challenges need to be considered when managing processes in cyber-physical domains. On the one hand, there is the need of representing explicitly real-world objects and "technical" aspects like device capability constraints, wireless networking, device mobility, etc. On the other hand, since cyber-physical domains are intrinsically "dynamic", a PMS that runs a process in such domains must be able to adapt itself to the current real world entities and environment.

To make our discussion more concrete, in Sect. 2.3.1 we present an application scenario (as running example) that comes from the emergency management domain and is inspired to a real disaster response plan investigated by the authors during the European project WORKPAD[1] [53–56]. Then, starting from the analysis of the application scenario and from the experience gained from participating to several European Projects involving CPSs, in Sect. 2.3.2 we identify a list of high-level features that a PMS aiming at managing and adapting processes in cyber-physical domains should provide.

2.3.1 A Running Example from the Emergency Management Domain

As an application scenario, let us consider the *emergency management* domain, in which teams of first responders act in disaster locations with the main purpose of assisting potential victims and stabilizing the situation. A CPS composed by first responders' mobile devices, robots and wireless communication technologies is coupled with a process-oriented approach for team coordination. A response plan encoded as a process and executed by a PMS deployed on mobile devices can help to coordinate the activities of first responders acting on the field.

To be more concrete, let us consider the emergency management situation described in Fig. 2.2a, in which a train derailment is depicted in a grid-type map. For the sake of simplicity, the train is composed of a locomotive (located at *loc*33) and two coaches (located at *loc*32 and *loc*31, respectively). In our train derailment situation, the goal of an incident response plan is to evacuate people from the coaches and take pictures for evaluating possible damages to the locomotive. To that end, a response team is sent to the derailment scene. The team is composed of four first responders, called *actors*, and two *robots*, initially all located at location cell *loc*00. It

---

[1]The WORKPAD Project (http://www.dis.uniroma1.it/~workpad) investigated how the use of a process-oriented approach can enhance the level of collaboration and support provided to first responders that act in emergency/disaster scenarios.

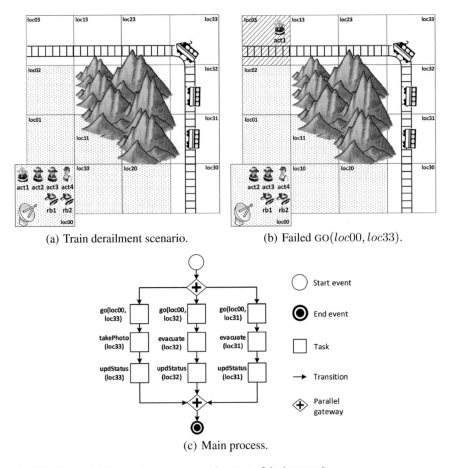

(a) Train derailment scenario.

(b) Failed GO($loc00, loc33$).

(c) Main process.

Fig. 2.2 A train derailment situation; area and context of the intervention

is assumed that actors are equipped with mobile devices for picking up and executing tasks, and that each provides specific capabilities. For example, actor $act1$ is able to extinguish fire and take pictures, while $act2$ and $act3$ can evacuate people from train coaches. The two robots, in turn, are designed to remove debris from specific locations. When the battery of a robot is discharged, actor $act4$ can charge it.

In order to carry on the response plan, all actors and robots ought to be continually inter-connected. The connection between mobile devices is supported by a fixed antenna located at $loc00$, whose range is limited to the dotted squares in Fig. 2.2a. Such a coverage can be extended by robots $rb1$ and $rb2$, which have their own independent (from antenna) connectivity to the network and can act as wireless routers to provide network connection in all adjacent locations. An incident response plan is defined by a set of activities that are meant to be executed on the field by first responders, and are predicated on specific contexts. Therefore, the information collected on-the-fly is used for defining and configuring at run-time the incident response plan at hand. A possible concrete realization of an incident response plan for our scenario is shown in Fig. 2.2c, using the BPMN modeling language. The process under investigation is composed of three parallel branches, with tasks instructing first responders to act for evacuating people from train coaches in $loc31$ and $loc32$, taking pictures of the locomotive, and assessing the gravity of the accident.

Due to the high dynamism of the environment, there are a wide range of exceptions that can ensue. Because of that, there is not a clear anticipated correlation between a change in the context and a change in the process. So, suppose for instance that actor $act1$ is sent to the locomotive's location, by assigning to it the task GO($loc00$, $loc33$) in the first parallel branch. Unfortunately, however, the actor happens to reach location $loc03$ instead. The actor is now located at a different position than the desired one, and most seriously, is out of the network connectivity range (cf. Fig. 2.2b). Since all participants need to be continually inter-connected to execute the process, the PMS has to first find a recovery procedure to bring back full connectivity, and then find a way to re-align the process. We notice that the execution of an emergency management process can also be jeopardized by the occurrence of *exogenous events* (e.g., a fire burnt up into a coach, a rock slide collapses in a location, etc.). Indeed, exogenous events could change, in asynchronous manner, some contextual properties of the scenario in which the process is under execution, hence possibly requiring the process to be adapted accordingly.

The above example (though it is very simple) shows that in a cyber-physical domain it is inadequate to assume that a process designer can pre-define all possible recovery activities for dealing with the exceptions that can ensue. The recovery procedures will depend on the actual context (e.g., the positions of process participants, the range of the main network, robot's battery levels, whether a location has become dangerous to get it, etc.) and there are too many of them to be considered at design-time. This emphasizes the fact that for processes enacted in cyber-physical domains there is a critical need of explicit mechanisms to model world changes and responding to them in a fully automated way.

2.3.2 High-Level Features for Managing Processes in Cyber-Physical Domains

The management of processes enacted in cyber-physical domains requires a PMS providing real-time monitoring and automated adaptation features during process execution [16]. To this end, the role of the data perspective becomes fundamental. Data, including information processes by process tasks as well as contextual information, is the main driver for triggering process adaptation, as focusing on the control flow perspective only would be insufficient. In fact, in a cyber-physical domain, a process is genuinely knowledge and data centric: the process control flow must be coupled with contextual data and knowledge production and process progression may be influenced by user decision making. This means that procedural and imperative models have to be extended and complemented with the introduction of declarative elements (e.g., tasks preconditions and effects) which enable a precise description of data elements and their relations, so as to go beyond simple process variables, and allow establishing a link between the control flow perspective and the data perspective.

Starting from the above considerations, coupled with the experience gained in the area and lessons learned from several European Projects involving CPSs (a partial list can be found in the Acknowledgements section at the end of the chapter), we derive a set of 5 high-level characterizing features that must be provided by a PMS that wants to successfully manage processes enacted in cyber-physical domains:

- [F1] *Representing digitally real-world objects.* The screening of real-world objects performed by the physical sensors disseminated in the world must be taken into consideration when planning and executing a collaborative process in cyber-physical domains. To make the PMS aware of physical reality, a *physical-to-digital bridge* that transforms the knowledge extracted from real-world objects in their digital counterpart is required.
- [F2] *Modeling contextual data.* Contextual data representing the cyber-physical domain in which the process will be enacted and all relevant data affecting the process and manipulated by it need to be formalized and encoded in an *information model*, so as to define data objects and information to be considered as part of the process context and execution state. A process designer should be also allowed to *express conditions over process data*, if needed.
- [F3] *Representing data-driven activities.* A process executed in a cyber-physical domain is characterized by activities whose enactment is related to the evolution of the information model. Such activities are *enriched with declarative elements and constraints* (e.g., preconditions and effects) defined on contextual data, which specify when a particular activity can be executed in a specific state of the contextual scenario, the execution dependencies between activities and the effects that activity executions have on the current state.
- [F4] *Monitoring and exception detection.* The PMS should automatically detect exceptional situations, i.e., any mismatch between the computerized version of the process and its corresponding real-world version. This requires to *monitor running process instances* against the evolution of the process execution context, to identify when a process instance is deviating from the intended behavior.
- [F5] *Exceptions resolution.* The PMS should react to any event that represents a risk for process continuity. If a detected anomalous situation may prevent process progressing, the PMS needs to *automatically deriving and enacting a recovery procedure* that allows the process to progress as expected.

If compared with the features provided by traditional control-flow oriented PMSs (a comprehensive list can be found in [57]), it is clear that processes enacted in cyber-physical domains reveal some challenging features (e.g., data orientation, low predictability, etc.) that pose serious problems for their support through the use of existing approaches [5]. While there is the lack of a holistic approach that allows to tackle the set of identified features as a whole and to provide a right support for them, we argue that the realization of such an approach can be regarded as a key success factor for the fruitful application of BPM in new domains different from the business one, and represents one main challenge that is currently under investigation by the research community [5, 16]. As a step towards this goal, in the following section we introduce the SmartPM approach and the corresponding implemented system. SmartPM provides a flexible approach to manage the life-cycle of processes enacted in cyber-physical domains, with a targeted support for the set of features discussed above.

2.4 The SmartPM Approach and System

SmartPM[2] (Smart Process Management) [58] is a model and a PMS implementing a set of techniques that enable automatically adapting process instances at run-time in the presence of unanticipated exceptions, without requiring an explicit definition of handlers/policies to recover from tasks failures and exogenous events, and without the intervention of domain experts at run-time.

The SmartPM approach builds on the dualism between an *expected reality*, the (idealized) model of reality that is used by the PMS to reason, and a *physical reality*, the real world with the actual values of conditions and outcomes. Process execution steps and exogenous events have an impact on the physical reality and any deviation from the expected reality results in a mismatch (or exception) to be removed to allow process progression. If an exception invalidates the enactment of the process being executed, an external state-of-the-art planner is invoked to synthesise a recovery procedure that adapts the faulty process instance by removing the gap between the two realities.

To meet the high-level features described in Sect. 2.3.2, SmartPM relies on and combines well-established AI techniques and frameworks, including the Situation Calculus [59], the IndiGolog framework [60] and automated planning [61]. The choice of adopting AI technologies is motivated by their ability to provide the right abstraction level needed when dealing with dynamic situations in which data (values) play a relevant role in system enactment and automated reasoning over the system progress. In the field of BPM, many other formalisms and technologies are being used, such as Petri Nets [62], Coloured Petri Nets [63], Workflow Nets [64], YAWL nets [30], BPMN [3] and process algebras [65], with varying degrees of automated reasoning support over them. While Petri Nets and Worklow Nets do not support data-based decisions as well as data-driven execution of any kind due to the lack of data-awareness [66], other formalisms such as Coloured Petri Nets, YAWL Nets, BPMN and Process Algebras are potentially all fine solutions for realizing our framework. However, the level of abstraction provided for manipulating data values and reasoning over dynamic changes is not formally specified (in the case of YAWL), performed at

---

[2]http://www.dis.uniroma1.it/~smartpm.

shallow level (in the case of BPMN) or at very low level (in the case of Coloured Petri Nets and Process Algebras), since such formalisms mainly focus on the control-flow perspective of a business process. Conversely, the AI field is rich of algorithms and systems that support the user in the creation, acquisition, adaptation, evolution, and sharing of data knowledge for specifying and implementing dynamic systems [59, 67, 68].

While the formal model underlying SmartPM is described in detail in [58], in this section we aim at providing an overview of the SmartPM approach (cf. Sect. 2.4.1), its concrete implementation (cf. Sect. 2.4.2) and application (cf. Sect. 2.4.3) to the running example introduced in Sect. 2.3.1.

2.4.1 Overview of the Approach

Process Representation

In SmartPM a process model includes a set T of n task definitions. Each task $t_i \in T$ is described in terms of its preconditions Pre_i and effects Eff_i, and can be considered as a single step that consumes input data and produces output data. Data are represented through a set F of *fluents* whose definition strictly depends on the specific process domain of interest. In AI, a fluent is a condition that can change over time. Such fluents can be used to constrain the task assignment (in terms of *task preconditions*), to assess the outcome of a task (in terms of *task effects*) and as guards for decision points and routing elements (e.g., for cycles or conditional statements).

SmartPM adopts a *service-based approach* to process management, that is, tasks are executed by services (that could be software applications, human actors, robots, agents, etc.). Choosing the fluents that are used to describe each activity falls into the general problem of *knowledge representation*. To this end, the environment, services and tasks are grounded in domain theories described in Situation Calculus [59]. Situation Calculus is specifically designed for representing dynamically changing worlds in which all changes are the result of task executions. Situation Calculus is thus used for providing a declarative specification of the domain (i.e., available tasks, contextual properties, tasks preconditions and effects, what is known about the initial state) where a process has to be executed. This declarative specification also covers the resource perspective, with a definition of the available services and the capabilities they provide, to be matched with capability requirements defined for the tasks.

On top of Situation Calculus, SmartPM relies on the IndiGolog high-level agent programming language for the specification of the process control flow. IndiGolog [60] enables the definition of programs with cycles, concurrency, conditional branching and interrupts that rely on program steps that are actions of some domain theory expressed in Situation Calculus. The dynamic world of SmartPM is modeled as progressing through a series of *situations*, where each situation s is the result of the

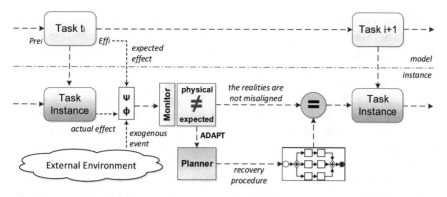

Fig. 2.3 Execution monitoring and adaptation in SmartPM

various tasks performed up to that point. In this context, fluents can be considered as "properties" of the world whose values may vary across situations.

Process Monitoring and Adaptation

SmartPM provides mechanisms for adapting process models that require no predefined handlers. To this end, a specialized version of the concept of adaptation from the field of agent-oriented programming [69] is used. The approach is schematized in Fig. 2.3. Specifically, adaptation in SmartPM can be seen as reducing the gap between the *expected reality*, i.e., the (idealized) model of reality that is used by the PMS to reason, and the *physical reality*, i.e., the real world with the actual values of conditions and outcomes.

The physical reality ϕ_s reflects the concept of "now", i.e., what is happening in the real environment while the process is under execution. The physical reality ϕ_s captures exactly the value assumed by each fluent in the situation s. In general, a task t_i can only be performed in a given physical reality ϕ_s if and only if that reality satisfies the preconditions Pre_i of that task. Moreover, each task has also a set of effects Eff_i that change the current physical reality ϕ_s into a new physical reality ϕ_{s+1}.

A PMS that takes as input a process specification should guarantee that each task is executed correctly, i.e., with an output that satisfies the process specification itself. In fact, at execution time, the process can be easily invalidated because of task failures or since the environment may change due to some external exogenous event. For this purpose, the concept of expected reality ψ_s is introduced. The expected reality in a situation s is given by the set of fluents that are supposed to hold. Basically, when a task is executed and completed, both the physical and expected realities are updated so that:

- the physical reality reflects the actual outcome produced by the task execution;
- the expected reality reflects the intended outcome of the task execution, according to the specification of task's effects.

A recovery procedure is needed in a specific situation if the two realities are different from each other. A misalignment of the two realities often stems from errors in the tasks outcomes (e.g., incorrect data values) or is the result of exogenous events coming from the environment. An execution monitor is responsible for detecting whether the gap between the expected and physical realities is such that the process instance cannot progress. In that case, the PMS has to find a recovery procedure whose execution removes the gap between the physical reality and the expected one.

SmartPM allows the synthesis of a recovery procedure at run-time by invoking an external state-of-the-art planner [61]. Given as the goal condition the process state reflecting the expected reality, the planner searches for a plan that may turn the physical reality into the expected reality. The recovery procedure will be built by composing tasks stored in a specific repository. The repository contains both tasks used for defining the specific process instance under execution and other tasks built on the same contextual scenario and possibly used in past executions of the process. If a recovery plan exists, it will be executed by SmartPM for adapting the faulty process instance.

2.4.2   The SmartPM Environment and Architecture

The concrete implementation of the SmartPM approach has required to cover the modeling, execution and monitoring stages of the process life-cycle and to make explicit the connection of implemented processes with the real-world objects of the cyber-physical domain of interest. To that end, as shown in Fig. 2.4, the architecture of the SmartPM system relies on five architectural layers.

Presentation Layer

The *Presentation Layer* provides a GUI-based tool called SmartPM Definition Tool (cf. Fig. 2.5), which assists the process designer in the definition of a process model at design-time. The SmartPM Definition Tool has been developed using the Java SE 7 Platform, and the JGraphX open source graphical library.[3] To define a process model with the SmartPM Definition Tool means (i) to build a *tasks repository*, (ii) to define the process *control flow* and (iii) to formalize the *contextual knowledge* of the cyber-physical domain in which the process will be enacted.

Contextual knowledge is represented as a *domain theory* that includes all the information of the application domain, such as the people/services that may be involved in performing the process, the exogenous events, the contextual data and so forth. Data are represented through some *atomic terms* that range over a set of *data objects*, defined over some *data types*. In short, a data object depicts an entity of interest (e.g., a location, a capability, a service, etc.), while each data type explicitly specifies the data objects that represent the domain of that type. Under this representation, possible values of a data type univocally identify data

---

[3]http://www.jgraph.com/.

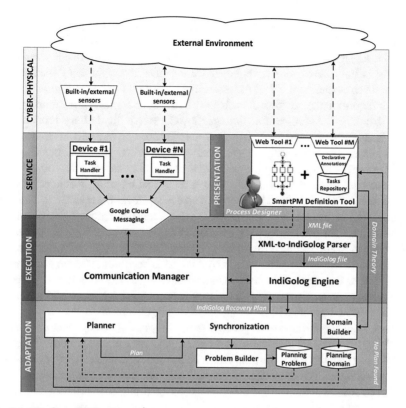

Fig. 2.4 The SmartPM architecture

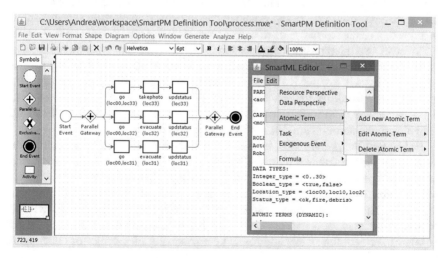

Fig. 2.5 A screenshot of the SmartPM Definition Tool

objects in the scenario of interest. *Atomic terms* can be used to express properties of domain objects (and relations over objects) and argument types of a term—taken from the set of the available data types—represent the finite domains over which the term is interpreted. For example, if we consider the emergency management domain discussed in Sect. 2.3.1, the term $At[act : Actor] = (loc : Location\_type)$ is used for recording the position of each actor in the area. Similarly, the numeric term $BatteryLevel[rbt : Robot] = (int : Integer\_type)$ records the battery level of each robot. In addition, the designer can define *complex terms*. They are declared as basic atomic terms, with the additional specification of a well-formed first-order formula that determines the truth value for the complex term. For example, the complex term $Connected[act : Actor]$ can be defined to express that an actor is connected to the network if s/he is in a covered location or if s/he is in a location adjacent to a location where a robot is located. For each atomic/complex term, the process designer has to decide which ones are *relevant* for adaptation and which ones have not to be considered for that. An atomic term that is considered as relevant for adaptation will be continuously monitored by the PMS, and if its value becomes different from the one expected, then the adaptation mechanisms provided by SmartPM will be triggered. A process designer can also specify which *exogenous events* may be catched at run-time and which atomic terms will be modified after their occurrence.

Concerning the definition of process *tasks*, the process designer is required to specify which tasks are applicable to the dynamic scenario under study. Tasks will be stored in a specific *tasks repository* and can be used for composing the control flow of the process and for adaptation purposes. Each task can be considered as a single step that consumes input data and produces output data, and is described with (i) typed input parameters, (ii) preconditions—defined over atomic and complex terms—that constrain the task assignment and must be satisfied before the task is applied, and (iii) deterministic effects, which establish the outcome of a task after its execution in terms of a change of the value of one or more atomic terms. For example, the task GO involves two input parameters *from* and *to* of type *Location_type*, representing a starting and an arrival location. An instance of this task can be executed only if the actor *SRV* that will execute it at run-time is at the starting location *from* and provides the required capabilities for executing the task. As a consequence of task execution, the actor moves from the starting to the arrival location, and this is reflected by assigning to the atomic term $At[SRV]$ the value *to* in the effect.

Notice that the definition of a valid domain theory and of tasks specifications allows to meet the features F2 and F3 introduced in Sect. 2.3.2. At this point, the process designer uses the BPMN graphical editor provided by the SmartPM Definition Tool to define the process control flow among a set of tasks selected from the tasks repository. The editor provides visual, graphical editing and creation of BPMN 2.0 business processes.[4] It is important to notice that atomic/complex terms can be used as guards for decision points and routing elements (e.g., for cycles or conditional

---

[4]The SmartPM Definition Tool provides a relevant subset of the BPMN modeling constructs to define the control flow of a process, including basic activities, start/end events and parallel/exclusive gateways.

statements). The outcome of the process design activity is a complete XML-encoded process specification that is passed to the *Execution Layer*.

Execution and Service Layers

The *Execution Layer* is in charge of managing and coordinating the process enactment. The BPMN process and the associated domain theory are taken as input from the *XML-to-IndiGolog Parser* component, a Java module that translates them into situation calculus [59] and IndiGolog [60] readable formats (cf. Sect. 2.4.1). It is interesting to notice that while from a user perspective the process control flow is defined using a subset of the modeling constructs provided by the BPMN notation, an *executable model* is obtained in the form of an IndiGolog program to be executed through an IndiGolog engine. To that end, we customized an existing IndiGolog engine,[5] written in the well-known open source SWI-Prolog environment,[6] to (i) build a physical reality by taking the initial context from the external environment; (ii) build an expected reality (initially equal to the physical one) that records the expected process state after each task execution or exogenous event occurrence; (iii) manage the process routing and decide which tasks are enabled for execution; (iv) collect exogenous events from the external environment. Once a task is ready for being executed, the IndiGolog engine is in charge of assigning it to a proper service (which may be a human actor, a robot, a software application, etc.) that is available (i.e., free from any other task assignment) and that provides all the required capabilities for task execution.

Process participants interact with the engine through a *Task Handler*, an interactive GUI-based software application that supports the visualization of assigned tasks and enables starting task execution and notifying of task completion by selecting an appropriate outcome (cf. Fig. 2.6a). The SmartPM Task Handler is realized for Android devices from version 4.0 and up. Each device has an unique ID that matches the service name defined in the domain theory by the designer. Every step of the task life cycle—ranging from the assignment to the release of a task—requires an interaction between the IndiGolog engine and the task handlers. Such an interaction is mainly intended for notifying the device corresponding to the human actor of actions performed by the IndiGolog engine as well as for notifying the engine of actions executed by actor through the task handler of the corresponding device. The communication between the IndiGolog engine and the task handler is mediated by the *Communicator Manager* component (which is essentially a web server) and established using the Google Cloud Messaging (GCM) service.[7]

As previously discussed, the IndiGolog engine is in charge of monitoring contextual data to identify changes or events which may affect process execution, and notify them to the *adaptation layer*. This allows to meet the feature F4 introduced in Sect. 2.3.2. Specifically, given a process instance δ, after each task completion (or exogenous event occurrence), the physical and expected realities are updated to reflect

---

[5]http://sourceforge.net/projects/indigolog/.

[6]http://www.swi-prolog.org/.

[7]https://developer.android.com/google/gcm/index.html.

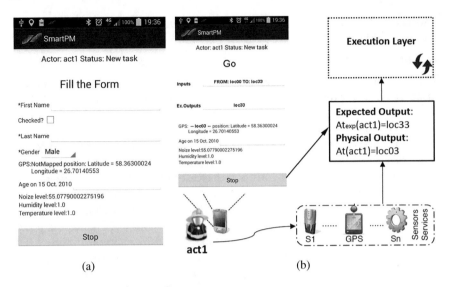

Fig. 2.6 The SmartPM Task Handler

the actual and intended (according to the specification of task's effects) outcome of task performance (or the contextual changes produced by an exogenous event). If we consider the first example shown in Sect. 2.3.1, when the task GO($loc00$, $loc33$) completes, it means that the output value for $At[act1]$ (generated as an effect of the task GO) is '$loc03$', that is different from the task's expected outcome, that is '$loc33$'. Hence, the two realities are misaligned, and the faulty process instance δ needs to be adapted (cf. Fig. 2.6b).

Adaptation Layer

To enable the automated synthesis of a recovery procedure, and to provide the right support to feature F5 discussed in Sect. 2.3.2, the *Adaptation Layer* of SmartPM resorts to classical planning techniques. Process adaptation relies on the capabilities provided by a PDDL-based planner component (the LPG-td planner [70]), which assumes the availability of a so-called *planning problem*, i.e., an initial state and a goal to be achieved, and of a *planning domain* definition that includes the actions to be composed to achieve the goal, the domain predicates and data types. To this end, if process adaptation is required, the *Domain Builder* component translates (i) the domain theory defined at design-time into a planning domain, while the *Problem Builder* component converts (ii) the physical reality into the initial state of the planning problem and (iii) the expected reality into the goal state of the planning problem. The planning domain and problem represent the input for the planner component and, in particular, the planning problem reflects the gap between the two realities. If the planner is able to synthesize a recovery procedure δ_a, the *Synchronization* component combines δ' (which is the remaining part of the faulty process instance δ still to be executed), with the recovery plan δ_a and builds an adapted process $\delta'' = (\delta_a; \delta')$. Notice that, whenever a process δ needs to be adapted, every running task is interrupted, since the recovery sequence of tasks δ_a has to be executed before that the remaining part of the process instance δ' can progress. Thus, active branches can only resume their execution after the repair sequence has been executed. This is fundamental to avoid the risk of introducing data inconsistencies during the repair phase. Finally, the *Synchronization* component converts δ'' into an executable IndiGolog program so that it can be enacted by the IndiGolog engine. Otherwise, if no plan exists for the current planning problem and no handling strategy can be automatically derived for the specific deviation, the control passes back to the process designer, who can try to manually manage the exception and adapt the process instance.

Cyber-Physical Layer

The cyber-physical layer is tightly coupled with the concrete physical components available in the cyber-physical domain under consideration. For automating the data collection from the environment, different built-in and external sensors can be used with the SmartPM Task Handler. To exploit sensors that are built in the mobile devices, several plugins have been created for the task handler. For example, location data can be obtained using built-in GPS sensors. Similarly, using the microphone, it is possible to automatically get the current noise level near the device. In addition, external sensors can be taken into use to gather automatic measurements—for prototyping purposes, the Arduino platform can be used.[8] The task handler can take advantage of this technology for gathering environmental data: Arduino has a large variety of sensors available to measure different environmental values, for example different gas levels in the air, water quality, radiation level, etc.; Arduino can be connected with Android via Bluetooth for transferring the data. We notice that the IndiGolog engine of SmartPM can only work with defined discrete values, while data gathered from physical sensors have naturally continuous values. Therefore, to meet feature F1 introduced in Sect. 2.3.2, a mapping of such continuous values into their discrete counterparts is required. To tackle this issue, we enhanced the SmartPM Definition Tool by providing several web tools that allow process designers to associate some of the data objects defined in the domain theory with the continuous data values collected from the environment. Notice that in SmartPM finiteness is crucial, as it is one of the main assumptions that makes classical planning possible to the computation of a recovery plan. For example, in the case of the GPS sensor, we developed a location web tool (as a Google Maps plugin) that allows a process designer to mark areas of interest from a real map (by selecting latidude/longitude values) and associate them to the discrete locations (e.g., *loc*00, *loc*01, etc.) defined during the design stage of a process through the SmartPM Definition Tool. Figure 2.7 shows a screenshot of the location web tool. Similarly, we developed further web tools for the other developed sensors (temperature, humidity, noise level, etc.). The mapping rules generated are then encoded in a XML file that is saved into the Communication Manager and retrieved at run-time (after any task completion) to allow the matching of the continuous data values collected by the specific sensor into discrete data objects.

---

[8]Arduino is an open-source physical computing platform based on a simple microcontroller board, and a development environment for writing software for the board, cf. http://arduino.cc/en/guide/introduction.

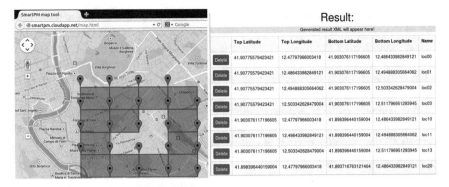

Fig. 2.7 A screenshot of the location web tool provided by the SmartPM Definition Tool

2.4.3 Applying SmartPM to the Running Example

While in the previous sections we discussed the approach underlying SmartPM and
the architecture of the implemented SmartPM system, in this section we present a
practical application of SmartPM with respect to the running example introduced in
Sect. 2.3.1. As anticipated in Sect. 2.4.2, the design of a process to be enacted in a
cyber-physical domain starts from the SmartPM Definition Tool, which supports the
process design activity. In SmartPM, the process design activity consists of defining
the domain theory, the tasks repository and the control flow of the process.

When defining a new domain theory, the very first step to perform involves specifying the *resource perspective* of the process, i.e., the *services* that will be involved
in tasks execution and the required *capabilities* to execute those tasks. If we consider
the emergency management scenario depicted in Sect. 2.3.1, the following services
and capabilities should be defined:

```
Service = {act1,act2,act3,act4,rb1,rb2}
Capability ={movement,hatchet,camera,gps,extinguisher,battery,
digger,powerpack}
```

We notice that the SmartPM Definition Tool allows to explicitly specify the *service
providers*, i.e., the real-world entities offering services to perform specific process
tasks. Examples of service providers are software components, smartphones, agents,
humans, robots, etc. In our running example, two kinds of providers are required,
actors and *robots*:

```
Actor = {act1,act2,act3,act4}
Robot = {rb1,rb2}
```

To make explicit which capabilities are provided by available services, a special
atomic term *Provides*[*srv* : *Service*, *cap* : *Capability*] (that is true if the capability
cap is provided by *srv* and false otherwise) is used. For example, to state that
actor *act*1 owns a mobile device with GPS capabilities, the term *Provides*[*act*1, *gps*]

will be set to `true`. Concerning the definition of data, two new data types (Boolean and Integer types are considered as predefined by the SmartPM Definition Tool) are required to capture the objects of interest in the emergency management domain under study.

```
Location_type = {loc00,loc10,loc20,loc30,loc01,loc11,loc02,loc03,
loc13,loc23,loc31,loc32,loc33}
Status_type = {ok,fire,debris}
```

The data type *Status_type* denotes the possible "states" of a location, while *Location_type* represents locations in the area. As discussed at the end of Sect. 2.4.2, data objects representing locations can be associated to real locations through a location web tool. The definition of data types and of the corresponding data objects allows the process designer to explicitly express the contextual properties of the cyber-physical domain under study. Such properties are captured through a finite number of *atomic* and *complex terms*. For our emergency management scenario, the following atomic and complex terms are required:

```
Evacuated[loc:Location_type]  = (bool:Boolean_type)
BatteryLevel[rbt:Robot] = (int:Integer_type)
PhotoTaken[loc:Location_type] = (int:Integer_type)
At[srv:Service]= (loc:Location_type)
Status[loc:Location_type] = (st:Status_type)
MoveStep[] = (int:Integer_type)
DebrisStep[] = (int:Integer_type)
Neigh[loc1:Location_type,loc2:Location_type] = (bool:Boolean_type)
Covered[loc:Location_type] = (bool:Boolean_type)
Connected[act:Actor] = {
EXISTS(l1:Location_type, l2:Location_type, rbt:Robot).
   ((at[act]=l1) AND (Covered[l1] OR (at[rbt]=l2 AND
Neigh[l1,l2]))))}
```

Therefore, we need boolean terms for indicating if people have been evacuated from a location (*Evacuated*), integer terms for representing the battery charge level of each robot (*BatteryLevel*) or for indicating the number of pictures taken in a specific location (*PhotoTaken*), and functional terms for recording the position of each actor/robot in the area (*At*) or for indicating if a specific location is safe, on fire or under debris (*Status*). Some atomic terms may be used as *constant values*. For example, the terms *MoveStep* and *DebrisStep* reflect the amount of battery consumed respectively when a robot moves from a location to another and when a robot removes debris from a specific location. Finally, atomic terms can also be used for expressing static relations over objects. For example, the atomic term *Neigh* indicates all adjacent locations in the area, while the atomic term *Covered* reflects the locations covered by the network provided by the fixed antenna. For each atomic term, the process designer may decide which ones are *relevant* for triggering the adaptation mechanisms provided by SmartPM. In our example, we can consider as relevant the atomic terms *At*, *Evacuated* and *PhotoTaken*. Finally, as anticipated in Sect. 2.4.2, our emergency management scenario requires also the definition of a complex term *Connected* to denote if an actor is connected to the network.

The definition of the domain theory is the basis to specify the tasks repository and the exogenous events required for the scenario under study. Our running example requires the following tasks and exogenous events:

```
Tasks Repository = {go, move, takephoto, evacuate, updstatus,
extinguishfire, chargebattery}
Ex_events = {photoLost, fireRisk, rockSlide}
```

For each task, the SmartPM Definition Tool provides a wizard-based editor to build a task specification and to define the single conditions composing the task precondi-tions and effects. We notice that the process designer is required to make explicit if a task effect can be considered as *supposed* or *automatic*. When a task returns some real-world outcome after its completion, we define that outcome as *supposed*, since its physical value may be different from the expected one as thought at design-time. This is the case, for example, of the effect of the task GO (the definition of the task GO has been provided in Sect. 2.4.2), whose consequence is to move an actor from a starting to an arrival location, which can be different from the one expected at design-time. Sometimes it may also happen that a task effect is *automatic*, i.e., it is applied every time a task completes its execution, independently from the outcomes returned by the task itself. For example, when a robot removes debris from a loca-tion, its battery decreases of a fixed quantity that does not depend on any physical outcome.

The procedure is similar for the definition of exogenous events. However, in this latter case there is no need to specify any precondition, while effects can only be con-sidered as *automatic* (i.e., they are automatically applied to the involved terms when the exogenous event is catched). For example, the exogenous event ROCKSLIDE(*loc*) alerts about a rock slide collapsed in location *loc*, and its effect changes the value of the atomic term *Status*[*loc*] to the value '*debris*'.

Starting from the domain theory and the tasks repository just defined, the control flow that captures the response plan of our running example can be built through the BPMN editor provided by the SmartPM Definition Tool (as shown in Fig. 2.5).

The very last step before executing the process consists of instantiating the domain theory with a *starting state*, which reflects an initial assignment of values to the atomic terms. This procedure is performed automatically by the SmartPM Definition Tool, which collects the values of the properties of the cyber-physical domain of interest by querying the sensors installed on services' devices. From a formal point of view, the definition of the starting state corresponds to the creation of the physical and expected realities. In the case of our running example, the initial physical and expected realities reflect the values of the contextual properties of the world before to execute any step of the emergency management process (cf. Fig. 2.2a). A fragment of two realities in the starting state S_0 is shown below:

- $\phi(S_0) = \{At[act1]=loc00, \dots, Connected[act1]=\text{true}, \dots, Status[loc31]=ok\}$
- $\psi(S_0) = \{At[act1]=loc00, \dots, Connected[act1]=\text{true}, \dots, Status[loc31]=ok\}$

During process enactment, SmartPM is in charge of assigning tasks to proper services and of continuously monitoring the evolution of the two realities. Let us consider again our running example, and suppose that actor *act*1 is sent to the loco-motive's location, by assigning to it the task GO(*loc*00, *loc*33) in the first parallel branch of the emergency management process defined in Fig. 2.2c. However, as depicted in Fig. 2.2b, the actor happens to reach location *loc*03 instead, meaning that it is now located at a different position than the desired one and is out of the network connectivity range. Consequently, the two realities change as follows:

- $\phi(S_1)$ = {*At*[*act*1]=*loc*03, ... , *Connected*[*act*1]=false, ... , *Status*[*loc*31]=*ok*}
- $\psi(S_1)$ = {*At*[*act*1]=*loc*33, ... , *Connected*[*act*1]=true, ... , *Status*[*loc*31]=*ok*}

To re-align the physical reality with the expected one, SmartPM has to first find a recovery procedure to bring back full connectivity, and then find a way to re-align the process. To that end, provided robots have enough battery charge, SmartPM may first instruct the first robot to move to cell *loc*03 (cf. Fig. 2.8a) in order to re-establish network connection to actor *act*1, and then instruct the second robot to reach location *loc*23 in order to extend the network range to cover the locomotive's location *loc*33. Finally, task GO(*loc*03, *loc*33) is reassigned to actor *act*1 (cf. Fig. 2.8b). The corresponding updated process is shown in Fig. 2.9a, with the encircled section being the recovery (adaptation) procedure. The two realities are updated as follows:

- $\phi(S_2)$ = {*At*[*act*1]=*loc*33, ... , *Connected*[*act*1]=true, ... , *Status*[*loc*31]=*ok*}
- $\psi(S_2)$ = {*At*[*act*1]=*loc*33, ... , *Connected*[*act*1]=true, ... , *Status*[*loc*31]=*ok*}

Notice that after the recovery procedure, the enactment of the original process can be resumed to its normal flow. For example, in the third parallel branch, actor *act*2 can now be instructed to reach *loc*31. However, even if *act*2 completes its task as expected (cf. Fig. 2.8c), a further exception is thrown. In fact, *act*2 is out of the network connectivity range and, again, the PMS may instruct the first robot to move from cell *loc*03 to cell *loc*20 in order to re-establish network connection to actor *act*2 (cf. top of Fig. 2.9b). At this point, *act*2 may start evacuating people from *loc*31.

As a further example, let us suppose now that a rock slide collapses in location *loc*31 (cf. Fig. 2.8c) while *act*2 is evaluating the damages in that area (i.e., *act*2 is executing the UPDATESTATUS(*loc*31) task). Such an exogenous event, which cor-responds to ROCKSLIDE(*loc*31), changes in asynchronous manner only the physical reality, as follows:

- $\phi(S_3)$ = {*At*[*act*1]=*loc*33, *Connected*[*act*1]=true , ..., *Status*[*loc*31]=*debris*}
- $\psi(S_3)$ = {*At*[*act*1]=*loc*33, *Connected*[*act*1]=true , ..., *Status*[*loc*31]=*ok*}

In such a case, SmartPM needs first to abort the running task UPDATESTATUS(*loc*31) (the presence of a rock slide may possibly prevent the correct execution of the task), and then to find a recovery procedure that allows to remove the rock slide from *loc*31 by maintaining all the process participants inter-connected to the network. A possible solution is shown in Fig. 2.8d, and consists of instructing *act*4 to reach *loc*20 for recharging the battery of *rb*1, of moving the robot *rb*1 in *loc*31 in order to remove debris, and finally of reassigning the UPDATESTATUS(*loc*31) task to *act*2. The corresponding adapted process is shown in the bottom of Fig. 2.9b, and the two realities are updated as follows:

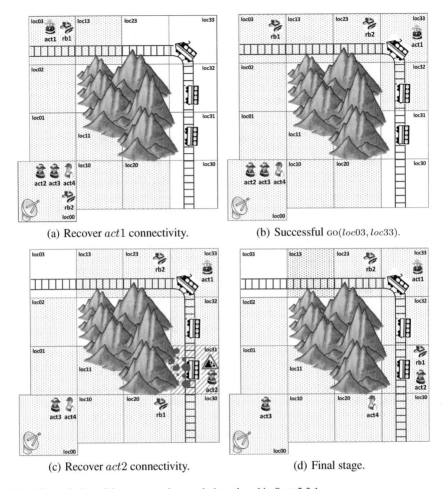

(a) Recover *act*1 connectivity.   (b) Successful GO(*loc*03, *loc*33).

(c) Recover *act*2 connectivity.   (d) Final stage.

Fig. 2.8 Evolution of the contextual scenario introduced in Sect. 2.3.1

- $\phi(S_4) = \{At[act1]=loc33,\ Connected[act1]=\texttt{true}\ ,\ ...,\ Status[loc31]=ok\}$
- $\psi(S_4) = \{At[act1]=loc33,\ Connected[act1]=\texttt{true}\ ,\ ...,\ Status[loc31]=ok\}$

It is worth noting that we validated the SmartPM approach with a case study based on real processes coming from the emergency management domain. Specifically, we first performed empirical experiments on synthetic data by enacting several emergency management processes, and they confirm the feasibility of the planning-based approach provided by SmartPM for adapting processes in medium-sized cyber-physical domains from the timing performance perspective. Then, we tested the SmartPM System with 3600 different process models having control flows with different structures (and different domain theories associated to them) to measure the effectiveness of SmartPM in adapting processes. We define the *effectiveness* of a PMS as the *ability of a PMS to complete the execution of a process model (i.e., to*

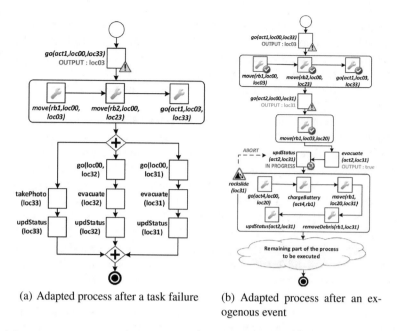

(a) Adapted process after a task failure   (b) Adapted process after an exogenous event

Fig. 2.9 Recovery procedures for the emergency response plan introduced in Sect. 2.3.1

execute all the tasks involved in a path from the start event to the end event) by adapting automatically its running process instance if some failure arises, without the need of any manual intervention of the process designer at run-time. To evaluate the effectiveness of SmartPM, we simulated processes execution by introducing task failures and exogenous events during the process enactment according to a given probability. As an instance, if the percentage of tasks failures was equal to 70 % and the process model to be executed was composed of 10 tasks, we had that 7 tasks of its running process instance completed with some physical outcome different from the one expected, thus requiring the process to be adapted. To sum up, SmartPM was able to complete 2537 process instances without any domain expert intervention, corresponding to an effectiveness of about 70.5%. For a detailed discussion of the above experiments, we invite the interested reader to refer to [58].

2.5 Discussion

The analysis performed in this chapter underlines that processes enacted in cyber-physical domains demand a more flexible approach to process management, recognizing the fact that in real-world environments process models quickly become outdated and hence require closer interweaving of modeling and execution. The fact is that the common strategy used by the adaptive PMSs to deal with exceptions is to *manually* or *semi-automatically* define recovery procedures at run-time. However, in cyber-physical domains, analyzing and defining these adaptations "manually" becomes time-demanding and error-prone. Indeed, the designer should have a global vision of the application and its context to define appropriate recovery actions, which becomes complicated when the number of relevant context features and their interleaving increases.

Conversely, the adaptation mechanism provided by SmartPM is based on execution monitoring for detecting failures and context changes, and allows to *automatically synthesize at run-time* the recovery procedures, without requiring to predefine any specific adaptation policy or exception handler at design-time. Furthermore, if compared with the existing techniques on process adaptation coming from the field of AI, the SmartPM approach provides unique features that make it particularly suitable for managing processes in cyber-physical domains. For example, if compared with the works [47, 48] (discussed in Sect. 2.2.3), the SmartPM approach adapts a running process instance by modifying only those parts of the process that need to be changed/adapted and keeps other parts stable. This is particularly important, as processes executed in cyber-physical domains often involve real human participants, and to completely re-define the process specification at run-time for adaptation purposes would mean to revolutionize the work-list of tasks assigned to the process participants. Finally, while closely related to works [49, 50], the SmartPM approach deals with changes in a more abstract and domain-independent way, by just checking misalignment between expected/physical realities. Conversely, the work [49] requires that the process designer explicitly defines the policies for detecting the exceptions at design-time, while the work [50] requires specification of a (domain-dependent) adaptation policy, based on volatile variables and when changes to them become relevant.

From a general perspective, the planning-based automated exception handling approach of SmartPM should be considered as complementary with respect to existing techniques, acting as a "bridge" between pre-planned approaches and unplanned approaches. When an exception occurs and is detected, the run-time engine may first check the availability of a predefined exception handler, and if no handler was defined, it can rely on an automated synthesis of the recovery process. In the case that a planning-based approach fails in synthesizing a suitable handler (or a handler is generated but its execution does not solve the exception), a human participant can be involved, leaving her/him the task of manually adapting the process instance.

The use of classical planning techniques for the synthesis of the recovery procedure has a twofold consequence. On the one hand, we can exploit the good performance of current state-of-the-art planners to solve medium-sized real-world problems as used in practice (cf. [58]). On the other hand, classical planning imposes some restrictions for addressing more expressive problems, including incomplete information, preferences and multiple task effects. To sum up, specific requirements frame the scope of applicability of the approach, which basically relies on the following assumptions:

1. process structure can be completely captured in a procedural predefined process model that explicitly defines the tasks and their execution constraints;
2. process execution context can be fully captured as part of the process model, i.e., complete information about a fully observable domain is available;
3. domain objects and contextual properties representing the state of the world can be reconducted to a finite set of finite-domain variables;
4. process tasks can be completely specified in term of I/O data elements, preconditions and deterministic effects.

Moreover, in addition to the full observability assumption, the approach relies on a high degree of *controllability* over the environment: when process execution deviates from the prespecified expected behavior (i.e., the physical reality deviates from the expected one), it should be possible to synthesize a recovery process whose execution modifies the environment (as reflected in the physical reality) so that the process instance can progress as expected, according to the prespecified model (basically, the physical reality is reconducted to the expected reality). When the operational environment and process state cannot be reconducted to their expected representation, we are back to the case where a process cannot be recovered to progress according to the predefined model, and it is the process itself that has to be (manually or semi-automatically) adapted to the changed environment.

   The above assumptions result from the need of balancing between modeling complexity and expressive power, and the practical requirements that enable exploiting classical planning tools. Although the need to explicitly model process execution context and annotate tasks with preconditions and effects may require some extra modeling effort at design-time (also considering that process modeling efforts are often mainly directed to the sole control flow perspective), the overhead is compensated at run-time by the possibility of automating exception handling procedures.

2.6 Conclusion

In this chapter we have introduced a general approach, a concrete framework and a PMS, called SmartPM, for automated adaptation of processes enacted in cyber-physical domains in case of unanticipated exceptions. The approach is based on declarative task specifications and planning techniques, and relies on the ability of automatically synthesizing recovery procedures at run-time. No predefined exception detection and handling logic is thus required. The current prototype of SmartPM is developed to be effectively used by process designers and practitioners. Users define processes in the well-known BPMN language, enriched with semantic annotations for expressing properties of tasks, which allow our interpreter to derive the IndiGolog program representing the process. Interfaces with human actors (as specific graphical user applications in Java) and software services (through Web service technologies) allow the core system to be effectively used for enacting processes.

Future work will include an extension of our approach to "stress" the above assumptions by making the approach applicable to less-controllable cyber-physical domains, such as *smart museums* and, in general, *smart spaces*. In fact, in the last years, the current widespread availability of wireless network technology for mass consumption has triggered the appearance of plenty of wireless and/or mobile devices providing applications able to enhance the visitors' experience in cultural sites. The "pre-fixed" and static visits of physical spaces have been turned into interactive dynamic experiences customized to the human visitors' behaviours and needs. In this context, a process can be used to personalize the visit of an individual into a smart space.

In addition, we aim at turning the *centralized control* provided by SmartPM (in which the reasoning is performed by a single entity, which subsequently instructs the process participants what to do) into a *decentralized control*, in which each participant will be provided with her/his mobile device with the SmartPM system installed into it. The challenge is to provide each SmartPM system with the ability to adapt the single processes of individual process participants by considering not only the local knowledge collected by the single participant, but also the knowledge produced by the other visitors of the smart space and the global knowledge provided by the smart space as a whole (e.g., the knowledge produced by the sensors installed in the smart space). As shown in [71, 72], our research is already going in this direction.

Acknowledgements This work has been partly supported over the years by the following projects: EU FP-6 WORKPAD, EU FP-7 SM4All, Italian Sapienza grant TESTMED, Italian Sapienza grant SUPER, Italian Sapienza award SPIRITLETS, Italian cluster Social Museum and Smart Tourism, Italian project NEPTIS, Italian project RoMA. The authors would like to thanks the many persons involved over the years in the SmartPM conception and development, namely Giuseppe De Giacomo, Massimiliano de Leoni, Patris Halapuu, Arthur H.M. ter Hofstede, Alessandro Russo, Sebastian Sardina, Paola Tucceri, Stefano Valentini.

References

1. Lee, E.A.: Cyber Physical Systems: Design Challenges. In: 11th IEEE International Symposium on Object-Oriented Real-Time Distributed Computing (ISORC). IEEE Computer Society, pp. 363–369 (2008)
2. Weske, M.: Business Process Management—Concepts, Languages, Architectures (2nd edn). Springer (2012)
3. Allweyer, T.: BPMN 2.0: Introduction to the Standard for Business Process Modeling. BoD–Books on Demand (2010)
4. Andrews, T., Curbera, F., Dholakia, H., Goland, Y., Klein, J., Leymann, F., Liu, K., Roller, D., Smith, D., Thatte, S., et al.: Business Process Execution Language for Web Services. Version 1.1. (2003)
5. Di Ciccio, C., Marrella, A., Russo, A.: Knowledge-intensive processes: characteristics, requirements and analysis of contemporary approaches. J. Data Semant. **4**(1), 29–57 (2015)
6. Di Ciccio, C., Marrella, A., Russo, A.: Knowledge-intensive processes: an overview of contemporary approaches. In: 1st International Workshop on Knowledge-intensive Business Processes (KiBP). CEUR Workshop Proceedings, vol. 861. CEUR-WS.org (2012)

7. Lenz, R., Reichert, M.: IT support for healthcare processes—premises, challenges, perspectives. Data Knowl. Eng. **61**(1), 39–58 (2007)
8. Cossu, F., Marrella, A., Mecella, M., Russo, A., Bertazzoni, G., Suppa, M., Grasso, F.: Improving operational support in hospital wards through vocal interfaces and process-awareness. In: 25th International Symposium on Computer-Based Medical Systems (CBMS), pp. 1–6. IEEE Computer Society (2012)
9. Cossu, F., Marrella, A., Mecella, M., Russo, A., Kimani, S., Bertazzoni, G., Colabianchi, A., Corona, A., Luise, A.D., Grasso, F., Suppa, M.: Supporting doctors through mobile multimodal interaction and process-aware execution of clinical guidelines. In: 7th International Conference on Service-Oriented Computing and Applications (SOCA), pp. 183–190. IEEE (2014)
10. Marrella, A., Mecella, M.: Continuous planning for solving business process adaptivity. In: 12th International Working Conference on Business Process Modeling, Development and Support (BPMDS). Lecture Notes in Business Information Processing, vol. 81, pp. 118–132. Springer (2011)
11. Marrella, A., Russo, A., Mecella, M.: Planlets: automatically recovering dynamic processes in YAWL. In: 20th International Conference on Cooperative Information Systems (CoopIS). Lecture Notes in Computer Science, vol. 7565, pp. 268–286. Springer (2012)
12. Seiger, R., Keller, C., Niebling, F., Schlegel, T.: Modelling complex and flexible processes for smart cyber-physical environments. J. Comput. Sci. (2014)
13. Helal, S., Mann, W., El-Zabadani, H., King, J., Kaddoura, Y., Jansen, E.: The gator tech smart house: a programmable pervasive space. IEEE Comput. **38**(3), 50–60 (2005)
14. Rajkumar, R.R., Lee, I., Sha, L., Stankovic, J.: Cyber-physical systems: the next computing revolution. In: 47th Design Automation Conference (DAC), pp. 731–736. ACM (2010)
15. Klein, M., Dellarocas, C.: A knowledge-based approach to handling exceptions in workflow systems. Comput. Support. Coop. Work (CSCW) **9**(3–4), 399–412 (2000)
16. Reichert, M., Weber, B.: Enabling Flexibility in Process-Aware Information Systems - Challenges, Methods, Technologies. Springer (2012)
17. Marrella, A., Mecella, M., Russo, A.: Featuring automatic adaptivity through workflow enactment and planning. In: 7th International Conference on Collaborative Computing: Networking, Applications and Worksharing (CollaborateCom), pp. 372–381. ICST/IEEE (2011)
18. Klein, M., Dellarocas, C., Bernstein, A.: Introduction to the special issue on adaptive workflow systems. Comput. Support. Coop. Work (CSCW) 9(3–4), 265–267 (2000)
19. Sadiq, S., Orlowska, M.: On capturing exceptions in workflow process models. In: 3rd International Conference on Business Information Systems (BIS), pp. 3–19. Springer London (2000)
20. Casati, F., Ceri, S., Paraboschi, S., Pozzi, G.: Specification and implementation of exceptions in workflow management systems. ACM Trans. Database Syst. (TODS) **24**(3), 405–451 (1999)
21. Casati, F., Cugola, G.: Error handling in process support systems. In: Advances in Exception Handling Techniques (ECOOP). Lecture Notes in Computer Science, vol. 2022, pp. 251–270. Springer (2001)
22. Eder, J., Liebhart, W.: The workflow activity model WAMO. In: Third International Conference on Cooperative Information Systems (CoopIS), pp. 87–98 (1995)
23. Eder, J., Liebhart, W.: Workflow recovery. In: First IFCIS International Conference on Cooperative Information Systems (CoopIS), pp. 124–134. IEEE Computer Society (1996)
24. Hagen, C., Alonso, G.: Exception handling in workflow management systems. IEEE Trans. Softw. Eng. **26**(10), 943–958 (2000)
25. Luo, Z., Sheth, A., Kochut, K., Miller, J.: Exception handling in workflow systems. Appl. Intell. **13**(2), 125–147 (2000)
26. Adams, M.J.: Facilitating Dynamic Flexibility and Exception Handling for Workflows. Ph.D. thesis, Queensland University of Technology Brisbane, Australia (2007)
27. Russell, N., van der Aalst, W.M.P., ter Hofstede, A.H.M.: Workflow exception patterns. Advanced Information Systems Engineering. Lecture Notes in Computer Science, vol. 4001, pp. 288–302. Springer, Berlin Heidelberg (2006)
28. Lerner, B.S., Christov, S., Osterweil, L.J., Bendraou, R., Kannengiesser, U., Wise, A.: Exception handling patterns for process modeling. IEEE Trans. Softw. Eng. **36**(2), 162–183 (2010)

29. Chiu, D.K.W., Li, Q., Karlapalem, K.: A logical framework for exception handling in ADOME workflow management system. In: 12th International Conference on Advanced Information Systems Engineering (CAiSE). Lecture Notes in Computer Science, vol. 1789, pp. 110–125. Springer (2000)

30. ter Hofstede, A.H.M., van der Aalst, W.M.P., Adams, M., Russell, N.: Modern Business Process Automation: YAWL and its Support Environment. Springer (2009)

31. Weske, M.: Formal foundation and conceptual design of dynamic adaptations in a workflow management system. In: 34th Annual Hawaii International Conference on System Sciences (HICSS). IEEE Computer Society (2001)

32. Weber, B., Reichert, M., Rinderle-Ma, S.: Change patterns and change support features—enhancing flexibility in process-aware information systems. Data Knowl. Eng. **66**(3), 438–466 (2008)

33. Reichert, M., Weber, B.: Process Change Patterns: Recent Research, Use Cases, Research Directions. In: Seminal Contributions to Information Systems Engineering, 25 Years of CAiSE, pp. 397–404. Springer (2013)

34. Rinderle, S., Reichert, M., Dadam, P.: Correctness criteria for dynamic changes in workflow systems—a survey. Data Knowl. Eng. **50**(1), 9–34 (2004)

35. Rinderle, S., Weber, B., Reichert, M., Wild, W.: Integrating process learning and process evolution—a semantics based approach. In: 3rd International Conference on Business Process Management (BPM). Springer (2005)

36. Weber, B., Wild, W., Breu, R.: CBRFlow: enabling adaptive workflow management through conversational case-based reasoning. Lecture Notes in Computer Science, vol. 3155, pp. 434–448. Springer (2004)

37. Minor, M., Bergmann, R., Görg, S.: Case-based adaptation of workflows. Inf. Syst. **40**, 142–152 (2014)

38. Reichert, M., Dadam, P.: ADEPTflex—supporting dynamic changes of workflows without losing control. J. Intell. Inf. Syst. **10**(2), 93–129 (1998)

39. Reichert, M., Rinderle, S., Dadam, P.: ADEPT workflow management system. In: Business Process Management (BPM). Lecture Notes in Computer Science, vol. 2678, pp. 370–379. Springer (2003)

40. Reichert, M., Rinderle, S., Kreher, U., Dadam, P.: Adaptive process management with ADEPT2. In: 21st International Conference on Data Engineering (ICDE), pp. 1113–1114. IEEE Computer Society (2005)

41. Lanz, A., Reichert, M., Dadam, P.: Robust and flexible error handling in the AristaFlow BPM suite. In: Information Systems Evolution—CAiSE Forum 2010, Selected Extended Paper. Lecture Notes in Business Information Processing, vol. 72, pp. 174–189. Springer (2011)

42. Müller, R., Greiner, U., Rahm, E.: AGENT WORK: a workflow system supporting rule-based workflow adaptation. Data Knowl. Eng. **51**(2) (2004)

43. Myers, K., Berry, P.: Workflow management systems: an AI perspective. AIC-SRI report (1998)

44. Beckstein, C., Klausner, J.: A meta level architecture for workflow management. J. Integr. Des. Process Sci. **3**(1), 15–26 (1999)

45. Jarvis, P., Moore, J., Stader, J., Macintosh, A., du Mont, A.C., Chung, P.: Exploiting AI technologies to realise adaptive workflow systems. AAAI Workshop on Agent-Based Systems in the Business Context (1999)

46. R-Moreno, M.D., Kearney, P.: Integrating AI planning techniques with workflow management system. Knowl. Based Syst. **15**(5–6), 285–291 (2002)

47. Gajewski, M., Meyer, H., Momotko, M., Schuschel, H., Weske, M.: Dynamic failure recovery of generated workflows. In: 16th International Workshop on Database and Expert Systems Applications (DEXA), pp. 982–986. IEEE Computer Society Press (2005)

48. Ferreira, H., Ferreira, D.: An integrated life cycle for workflow management based on learning and planning. Int. J. Coop. Inf. Syst. **15**(4), 485–505 (2006)

49. Bucchiarone, A., Pistore, M., Raik, H., Kazhamiakin, R.: Adaptation of service-based business processes by context-aware replanning. In: 4th International Conference on Service-Oriented Computing and Applications (SOCA), pp. 1–8 (2011)

50. van Beest, N., Kaldeli, E., Bulanov, P., Wortmann, J., Lazovik, A.: Automated runtime repair of business processes. Inf. Syst. **39**, 45–79 (2014)
51. Baheti, R., Gill, H.: Cyber-Physical Systems. The Impact of Control Technology. Technical Report (2011)
52. Neyem, A., Franco, D., Ochoa, S.F., Pino, J.A.: An approach to enable workflow in mobile work scenarios. In: 11th International Conference on Computer Supported Cooperative Work in Design IV (CSCWD), Lecture Notes in Computer Science, vol. 5236, pp. 498–509. Springer (2007)
53. Catarci, T., de Leoni, M., Marrella, A., Mecella, M., Salvatore, B., Vetere, G., Dustdar, S., Juszczyk, L., Manzoor, A., Truong, H.L.: Pervasive software environments for supporting disaster responses. IEEE Internet Comput. **12**(1), 26–37 (2008)
54. Humayoun, S.R., Catarci, T., de Leoni, M., Marrella, A., Mecella, M., Bortenschlager, M., Steinmann, R.: The WORKPAD user interface and methodology: developing smart and effective mobile applications for emergency operators. In: 5th International Conference on Universal Access in Human-Computer Interaction. Applications and Services (UAHCI). Lecture Notes in Computer Science, vol. 5616, pp. 343–352. Springer (2009)
55. Catarci, T., de Leoni, M., Marrella, A., Mecella, M., Russo, A., Steinmann, R., Bortenschlager, M.: WORKPAD: process management and geo-collaboration help disaster response. Int. J. Inf. Syst. Crisis Response Manag. (IJISCRAM) **3**(1), 32–49 (2011)
56. Marrella, A., Mecella, M., Russo, A.: Collaboration on-the-field: suggestions and beyond. In: 8th International Conference on Information Systems for Crisis Response and Management (ISCRAM) (2011)
57. van der Aalst, W.M.P.: Business process management: a comprehensive survey. ISRN Software Engineering (2013)
58. Marrella, A., Mecella, M., Sardina, S.: SmartPM: An adaptive process management system through situation calculus, indigolog, and classical planning. In: 14th International Conference on Principles of Knowledge Representation and Reasoning (KR). AAAI Press (2014)
59. Reiter, R.: Knowledge in Action: Logical Foundations for Specifying and Implementing Dynamical Systems. MIT Press (2001)
60. De Giacomo, G., Lespérance, Y., Levesque, H., Sardina, S.: IndiGolog: a high-level programming language for embedded reasoning agents. In: Multi-Agent Programming, pp. 31–72. Springer, US (2009)
61. Nau, D., Ghallab, M., Traverso, P.: Autom. Plan. Theory Pract. Morgan Kaufmann Publishers Inc., San Francisco, CA, USA (2004)
62. van Der Aalst, W.M.P.: Three good reasons for using a petri-net-based workflow management system. In: International Working Conference on Information and Process Integration in Enterprises (IPIC'96), pp. 179–201. Cambridge, Massachusetts (1996)
63. Jensen, K.: Coloured Petri Nets. In: Petri Nets: Central Models and their Properties, pp. 248–299. Springer (1987)
64. van der Aalst, W.M.P.: The application of petri nets to workflow management. J. Circuits Syst. Comput. **8**(01), 21–66 (1998)
65. Puhlmann, F., Weske, M.: Using the π-calculus for formalizing workflow patterns. Business Process Management (BPM). Lecture Notes in Computer Science, vol. 3649, pp. 153–168. Springer, Berlin Heidelberg (2005)
66. Meyer, A., Smirnov, S., Weske, M.: Data in Business Processes. No. 50, Universitätsverlag Potsdam (2011)
67. Reichgelt, H.: Knowledge representation: an AI perspective. Ablex (1991)
68. Brachman, R., Levesque, H.: Knowledge Representation and Reasoning. Morgan Kaufmann Publishers Inc. (2004)
69. De Giacomo, G., Reiter, R., Soutchanski, M.: Execution monitoring of high-level robot programs. In: 6th International Conference on Principles of Knowledge Representation and Reasoning (KR), pp. 453–465. Morgan Kaufmann (1998)
70. Gerevini, A., Saetti, A., Serina, I., Toninelli, P.: LPG-TD: a fully automated planner for PDDL2.2 domains. In: 14th International Conference on Automated Planning and Scheduling (ICAPS). International Planning Competition abstracts (2004)

71. Marrella, A., Lespérance, Y.: Towards a Goal-oriented framework for the automatic synthesis of underspecified activities in dynamic processes. In: 6th International Conference on Service-Oriented Computing and Applications (SOCA), pp. 361–365. IEEE (2013)
72. Marrella, A., Lespérance, Y.: Synthesizing a library of process templates through partial-order planning algorithms. In: 14th International Working Conference on Business Process Modeling, Development and Support (BPMDS). Lecture Notes in Business Information Processing, vol. 147, pp. 277–291. Springer (2013)

Chapter 3
Towards Executable Specifications for Case Management Processes

Irina Rychkova, Bénédicte Le Grand and Carine Souveyet

Abstract Explicit process specifications play an important role in process-aware information systems (PAIS). Whereas methodologies for modeling structured, activity-oriented processes are well established, modeling formalisms for unstructured processes such as case management processes (CMP) are lagging. In this chapter, we define a state-oriented formalism that allows for executable specifications of CMP and paves the road for predictive analysis and recommendations support intended to case managers. This formalism is grounded on statecharts developed by D. Harel in 1987. We adopt the main concepts defined by statecharts and demonstrate how they can be used to specify a case management process. We also propose adaptations and potential extensions of the statecharts formalism that could address CMP specifics and complexity.

Keywords Case management · Simulation-based testing · Automated process analysis · Recommendations · State machines · Statecharts

3.1 Introduction

A Process-Aware Information System (PAIS) is a software system that manages and executes operational processes involving people, applications, and/or information sources on the basis of process models [16]. Workflow management systems and BPM systems are classic examples of PAIS.

I. Rychkova (✉) · B. Le Grand · C. Souveyet
Université Paris 1 Panthéon-Sorbonne, 12, Place du Panthéon, 75005 Paris, France
e-mail: irina.rychkova@univ-paris1.fr
URL: http://www.univ-paris1.fr/centres-de-recherche/cri/

B. Le Grand
e-mail: benedicte.le-grand@univ-paris1.fr

C. Souveyet
e-mail: carine.souveyet@univ-paris1.fr

© Springer International Publishing AG 2017
G. Grambow et al. (eds.), *Advances in Intelligent Process-Aware
Information Systems*, Intelligent Systems Reference Library 123,
DOI 10.1007/978-3-319-52181-7_3

49

Started by F. Taylor and H. Ford, a pursuit of process optimization and automation resulted in the creation of workflow concepts, where a process is specified with a (predefined) flow of tasks [55]. Workflows provide a powerful formalism for the design, simulation, analysis as well as management and execution of *structured, activity-oriented processes*.

Today, practitioners express the increasing need for information systems supporting *unstructured, data-oriented processes* such as *case management processes (CMP)*. The Object Management Group (OMG) defines case management as "a coordinative and goal-oriented discipline, to handle cases from opening to closure, interactively between persons involved with the subject of the case and a case manager or case team" [34]. Davenport [12] defines a case management process as a process that is not predefined or repeatable, but instead, depends on its evolving circumstances and on decisions regarding a particular situation, i.e., a case. Claim processing, residence permit issuing, crisis management, and organization of events are examples of CMP.

PAIS supporting case management are gaining momentum nowadays. Among successful solutions the IBM Advanced Case Manager,[1] ISIS Papyrus,[2] Computas[3] or IBM Intelligent Operations Center[4] can be cited. Many solutions supporting case management are now being developed and reported by the community of practitioners promoting Adaptive Case Management (ACM) [49].

Explicit process specifications play an important role in PAIS: they allow for better communication between stakeholders, enable process analysis and support redesign efforts [2]. Methodologies, specification languages and environments for workflow modeling and analysis are widely presented in the literature and recognised by practitioners. In contrast, current CMP supporting solutions are mostly focused on process configuration and execution. Very little support for CMP modeling and analysis is provided.

In this chapter, we define a state-oriented formalism for the incremental and interactive modeling and simulation of CMP. Our formalism is grounded on statecharts developed by D. Harel in 1987 [19]. In particular, we explain (a) why statecharts is a suitable formalism for CMP, (b) how statecharts can be adOpted and adApted for specifying CMP; we also show (c) how executable statecharts specifications can be used for CMP simulation and (d) how they can enable predictive analysis and recommendation support for a case manager.

Statecharts were originally created as a visual, fully executable formalism for the specification, design and analysis of complex discrete-event systems. Case management processes share a number of characteristics with complex discrete-event systems [19, 20, 23]: they continuously interact with their environment, respond to

---

[1]http://www-03.ibm.com/software/products/en/category/advanced-case-management.

[2]http://www.isis-papyrus.com/.

[3]http://www.computas.com/.

[4]http://www-03.ibm.com/software/products/en/intelligent-operations-center.

unexpected events (interrupts) and have many possible operation scenarios. In particular, a CMP can be compared to a reactive system, for which the main challenge is to identify the appropriate activity or group of activities to perform in reaction to a given internal or external stimulus in a given situation (context). However, contrary to conventional reactive systems, CMP has a *goal* that can be reached by several alternative scenarios. Moreover, decisions about these scenarios in CMP are typically made by a human actor (the case manager). Therefore, a CMP supporting system can seldom automatically execute the activities but it can enable or recommend them for execution.

The statecharts formalism combines an intuitive and concise visual notation with precise semantics [21, 31]. Rhapsody [20] (now IBM Rational Rhapsody[5]) and the open source YAKINDU Statechart Tools (SCT)[6] are examples of statecharts modeling environments, where visual statecharts specifications can be created and executed.

Following the points stated above, *we adopt the main concepts of statecharts*, such as states and state hierarchies, transitions, triggering events, concurrency and broadcast communication for CMP specification. In order to address CMP specific features, *we extend the statecharts formalism* with the notions of goal and path; we also revisit the semantics behind triggering events and introduce the concept of event duration.

The advantages of statecharts specifications can be perceived both during the design of CMP and during their execution. As we will explain in this chapter:

- Statecharts specifications allow for incremental CMP design;
- Executable statecharts specification can be used for the simulation-based testing of CMP scenarios;
- Executable statechart specifications pave the road for automated recommendations for CMP.

We apply the proposed formalism to specify an example of CMP: a crisis (flood) management process defined for Hauts-de-Seine department of France.

The remainder of this chapter is organized as follows. In Sect. 3.2, we introduce our example and provide the terminology that will be used in this chapter. This terminology spans across two domains: complex systems and case management. In Sect. 3.3, we present and discuss the related work in CMP management and modeling. In Sect. 3.4, we introduce the statecharts formalism and draw the parallels between complex discrete-event systems and CMP. In Sect. 3.5, we demonstrate how the statecharts formalism can be adopted and extended in order to provide fully executable specifications of CMP. In Sect. 3.6, we discuss the prospective added value of executable CMP specifications, trace a roadmap for future research and draw our conclusions.

---

[5]http://www-03.ibm.com/software/products/en/ratirhapfami.
[6]http://statecharts.org/index.html.

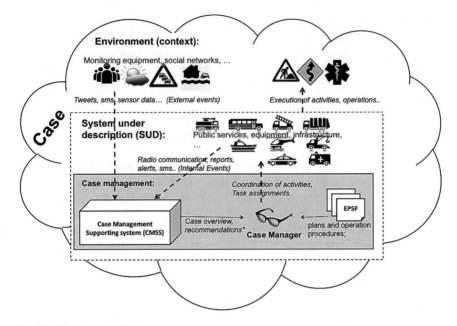

Fig. 3.1 The scope the flood management process

3.2 Case Management Process Example and Terminology

In this section we provide an example of CMP—a crisis management process designed to handle floods (we will call it flood management process) in a French department Hauts-de-Seine. We also briefly introduce the terminology used in this chapter and illustrate it on our example. Figure 3.1 shows the scope of the flood management process.

3.2.1 Crisis Management in Cases of Flood

A *flood* is an overflow of water that submerges a land. It happens, for example, because of an increase in the flow of a river provoked by significant rainfalls. The risk of a "major flood" is the main natural risk in the Ile-de-France region, particularly during the winter period from November to March. Cities like Paris[7] are confronted to this risk: if a flood occurs, important damages can be expected, affecting thousands of people. Floods are considered harmful when the water level of the Seine river

---

[7]See http://cartorisque.prim.net/dpt/75/75_ip.html.

exceeds 5.50 m according to the scale on the Austerlitz bridge in Paris. In the Hauts-de-Seine department, the risk of flood is considered as particularly important since 1910.[8]

The goal of the flood management process is to maintain the proper operation of city infrastructure (water supply, electricity, telecommunication, road networks, public transport and so on) and to protect people and facilities from flood consequences. This process is a typical example of CMP:

- it demands interaction between multiple actors (government, public services, volunteers, etc.).
- it is driven by the dynamic context of the case (i.e., flood development, current status of vulnerable areas and of rescuing operations) rather than by a predefined sequence of activities.

Flood Emergency begins when the water level rises above 5.5 m at the Austerlitz Bridge and is supposed to keep rising (according to weather forecasts). At this stage, the centers for crisis management are set up and the Emergency Plan Specialized on Floods (EPSF) is triggered. The city services (rescue, fire fighters, police, etc.) therefore carry out specific activities accordingly.

The regional authorities monitor the crisis situation and coordinate the operation procedures in the following major areas: *evacuation of population and facilities, temporary accommodation, public transport, road traffic, water supply, electricity supply and telecommunications.*

According to the flood severity, the EPSF identifies different phases of flood emergency for each of these seven areas and specifies the procedures to control the situation and to protect the population and facilities.

For example, when the water level exceeds 6.25 m, the drinking water supply is reduced for the towns of Saint-Cloud, Garches, Vaucresson, Marnes la Coquette and Ville d'Avray. When the water level reaches 6.7 m, the drinking water supply for these towns is completely disrupted. Therefore, the provisioning and distribution of bottled drinking water should start as soon as the water level at Austerlitz Bridge reaches 6.25 m. In case of limited supply, prioritized water provisioning has to be organized.

Along those lines, depending on the water level, various procedures are launched: a partial or complete interruption of public transport (SNCF Paris Rive Gauche, RER C, RATP), deviation and blocking of main highways (A86, A14, N14, etc.), evacuation of people, health care and childcare facilities.

Resources available for crisis management also need to be constantly monitored. In case of deficiencies in equipment, manpower or other problems that can compromise the crisis handling in one or several areas, specific measures such as mobilization of volunteers or federal alert raising can be taken. We model the resources as a specific area of the EPSF.

---

[8] Source: Préfecture des Hauts-de-Seine: Plan de secours spécialisé sur les inondations Hauts-de-Seine, SIDPC 21/11/2005, (2005), Available at: http://www.ville-neuillysurseine.fr/files/neuilly/mairie/services_techniques/plan-secours-inondation.pdf.

3.2.2   Terminology Used in This Chapter

A *case* is a situation (e.g., a flood crisis), which requires resolution. It is described by a set of elements that are relevant to or involved in a CMP. Within the case, we define the system boundary and distinguish between so-called system elements and context elements (that belong to the environment):

Case = System under description + Environment

The *System Under Description (SUD)* is described by the set of elements that can be controlled during the case management: public services, equipment, infrastructure, administration etc. It also includes a Case Management Supporting System (CMSS) and a case manager.

The SUD reacts to various stimuli (events) produced by the environment (e.g., change in temperature, water level, incidents) and performs activities in order to maintain the functioning of city infrastructure and to protect people and facilities from flood consequences.

The SUD produces internal events such as messages, reports and alerts sent by the agents via radio or mobile network. They can indicate the success or failure of a mission, resource deficiencies, emergency situations and so on.

The *environment* is described by the set of elements that interact with the SUD. It cannot be controlled but only monitored using specific equipment (e.g., meteo stations for monitoring weather, embedded sensors for measuring water level, video cameras for measuring traffic, social networks for collecting information about areas affected by the flood). The environment's behavior is unpredictable and brings uncertainty in the CMP.

The environment produces external events such as accidents, traffic jams, electric outages, malfunctioning of telecommunication.

The *Case Management Process (CMP)* describes the behavior of the SUD and defines what it has to do in order to achieve some objectives, i.e., to ensure safety and security for people and goods during the flood, until the emergency is over.

The *case management* element in Fig. 3.1 depicts a subsystem of the SUD which is responsible for the coordination of SUDs activities. It includes the case manager and the case management supporting system (CMSS):

The *Case Management Supporting System (CMSS)* is a PAIS for case management. The *case manager* is a human actor who uses the CMSS in order to monitor the case, to take decisions regarding the case handling scenario and to coordinate the activities of the SUD.

3.3   Related Work

In this section, we discuss Adaptive Case Management—for now, the most prominent paradigm for CMP support. We also review the existing modeling paradigms and formalisms for process specification and their capacity to model CMP.

3.3.1 Adaptive Case Management

The concept of Adaptive Case Management (ACM) has been defined as an "information technology that exposes structured and unstructured business information (business data and content) and allows structured (business) and unstructured (social) organizations to execute work (routine and emergent processes) in a secure but transparent manner".[9]

One of the major challenges identified by the ACM community, is the attempt to deal with CMP in the industry the same way as with regular business process—i.e., representing a case management by a workflow and focusing on the (predefined) *sequence of tasks*. This view implies that the data emerges and evolves within a process according to a predefined control flow similarly to a product evolving on a conveyor belt.

According to ACM [52], CMP must be organized around *a collection of data artifacts* about the case; the tasks and their ordering shall be adapted at run time, according to the evolution of the case circumstances and case-related data [41].

The body of knowledge on ACM has been extensively developed by practitioners; the best solutions are regularly reported in the book series on WfMC Global Awards for Excellence in Case Management [53, 54]. However, methodologies and formalisms for CMP modeling are rarely discussed.

3.3.2 Modeling Paradigms for CMP Specification

The important role of modeling in PAIS is discussed in [2]. The following general process modeling paradigms are identified in the literature [10, 11, 15]: activity-oriented, product (or state)-oriented and decision (or goal)-oriented.

The choice of a modeling paradigm depends on the conceptual properties of the process (e.g., flexibility vs. control).

According to the literature, case management processes (CMP) have the following conceptual properties:

1. CMP are unstructured, with non-repeatable execution scenarios [9, 52];
2. CMP are data-centered and are organized around a collection of data artifacts about the case [6, 52];
3. CMP are reactive and event-driven: activities should be carried out in reaction to a given internal or external event;
4. CMP must be considered within their context and the boundary between the system and its environment and the scope of the process must be clearly specified [6, 27, 48];
5. CMP are goal-oriented and flexible: goals are set and can be modified, added or removed during the execution [48];

---

[9]http://www.xpdl.org/nugen/p/adaptive-case-management/public.htm.

6. CMP are knowledge-intensive: decisions about the process scenario are made by a human actor—a knowledge worker—and are based on her knowledge, experience and intuition [25, 41];
7. CMP are unpredictable—they have to deal with events and handle the situations that were not planned or even imagined before [52].

In this section, we discuss the capacity of activity, product (state) and goal-oriented paradigms to express these conceptual properties of CMP.

Within the *activity-oriented paradigm*, the process is specified as an ordered set of activities that the system has to carry out. Examples of activity-oriented formalisms include BPMN [35], YAWL [3], activity diagrams in UML [46] and other languages based on workflow concepts.

Activity-oriented process modeling implies that data emerges and evolves within a process according to a predefined control flow. Events are supposed to occur (or be processed) at specific moments of the execution predefined by the model. This paradigm suits predictable and highly repeatable processes. CMP are unpredictable processes [52]: events and process inputs can occur at any time during execution; the order of activities cannot be predefined and depends on the current situation. Such behavior can therefore not be captured by the workflow formalism.

In order to increase process flexibility and to better address unstructured and knowledge-intensive processes like CMP, activity-oriented formalisms are extended with declarative parts, such as constraints [5], business rules [7] or configurable elements [45]. These formalisms can handle process variability within a potentially large number of configurations or scenarios. However, either such scenarios must be well identified upfront or the set of business rules (or configuration elements) must be regularly maintained by an expert. This can be seen as a limitation for CMP.

Techniques and frameworks for the analysis of activity-oriented process models are widely presented in the literature [57]. To provide automated process analysis, activity-oriented modeling languages are often annotated with or translated to some formal specification languages. The Declare framework [37] is a constraint-based system that uses a declarative language grounded on temporal logics. In [1], the state-oriented formalism of Petri Nets is used for workflow specification and analysis. In [14], the Petri Nets semantics for BPMN is presented. In [28], a business process model is mapped into a nondeterministic state machine for further analysis.

According to the *product-oriented (or state-oriented) paradigm*, a process is seen as a product life cycle (a set of product states and transitions between these states). Examples of product-oriented modeling formalisms include statemachines in UML [20], generic state-transition systems or state machines, such as FSM [38] or Petri Nets [32], and statecharts by D. Harel [19] created for the specification and analysis of complex discrete-event systems.

Within this paradigm, carried out activities depend on the current state of the product and the process scenario is adapted at run time, according to the evolution of the product. This paradigm suits well reactive systems specification [23] since the system's response to an event shall be defined not only by the type of this event but also by the current situation of the system i.e., its state.

Several research groups are reporting on approaches to design and specification of unstructured, knowledge-intensive processes (including CMP) based on the product-oriented paradigm.

In [9], process instances are represented as moving through a state space, and the process model is represented as a set of formal rules describing valid trajectories. Compared to our proposal based on statecharts, this approach is grounded on the theory of automated control systems. In [24], a group of researchers from IBM incorporates process- and data-centered perspectives; their approach is based on the concept of business artifacts. The Case Management Model and Notation is presented in [36]. This specification "is intended to capture the common elements that Case management products use, while also taking into account current research contributions on Case management." In [42], the Product-Based Workflow Design is presented. This approach explores the interaction between a product data model that reflects the product design and the process to manufacture this product represented by a workflow. The authors of [6] present case handling as a paradigm for supporting knowledge-intensive business processes. They recognise the lack of flexibility of workflow management systems and acknowledge the important role played by the "product"—the case—in the case handling. Their view on the case, however, remains activity-oriented: the proposed case definition explicitly includes the list of activities and their precedence relation assuming that they are known in advance.

Formalisms based on state machines are suitable for automated analysis including simulation, formal validation and model checking. Algorithms from graph theory can be used in order to analyse states reachability, "dead" states, path search and optimisation (where the path represents a process execution scenario).

In [24], the operational semantics of Guard-Stage-Milestone is presented. This semantics explains the interactions between business artifacts which are formalized following declarative principles. In our earlier work [47], we define formal semantics for CMP using the Alloy specification language. The Alloy Analyzer tool allows us to simulate and validate a CMP model; it also provides visual diagrams. Compared to statecharts, however, Alloy model is difficult to construct.

The product-oriented paradigm seems to be a good choice for specifying CMP. However, it does not support decision making since it does not define a notion of objective or goal.

The *decision or goal-oriented paradigm* extends the product-oriented view on the process: the successive transformations of the product are looked upon as consequences of decisions leading to some goal [33].

Goal-oriented modeling formalisms support decision making by specifying goal hierarchies and tracing each decision within these hierarchies. The examples include i∗[58], KAOS [30], MAP [43].

Goal-oriented formalisms extended with the notion of context are presented in [40, 44, 51]. These formalisms link a decision (expressed as a goal) to the situation in which this decision is taken (product state): for each state, a set of achievable and non-achievable goals can be identified and vice versa, each goal can be expressed in terms of states that the product has to reach. These formalisms can also connect the goals and the activities that must/can be carried out in order to achieve these goals.

The Generic Process Model (GPM) [51] is an example of context-driven goal-oriented formalism. It captures the process context and allows for reasoning about process goals. It is also suitable for automated process analysis.

Context-driven goal-oriented process models support automated recommendations and user guidance, providing that for each goal all the situations (states) in which this goal is achievable are known. Due to unpredictable sequences of events and non-repeatable execution scenarios in CMP, however, it will be hard if at all possible to model relations between various process situations, goals and activities that must/can be executed in order to achieve these goals. Such relations can be, though, discovered using process mining techniques (this is an interesting subject that lies behind the scope of this work).

Our analysis of existing modeling paradigms and their corresponding formalisms shows that the activity-oriented paradigm can hardly provide the flexibility required by CMP as expressed by their conceptual properties 1–3 and 5–7 listed above. Configurable specifications and business rules can be used to overcome the rigidity of traditional workflow-based formalisms, addressing properties 5 and 1–3 respectively. Nevertheless, they support the variability of process scenarios only within some boundaries defined by a number of business rules or configurable elements. Thus, they fail to address properties 6 and 7 of CMP.

The goal-oriented paradigm offers flexibility and supports knowledge workers. However, goal-modeling formalisms are typically suitable for an early phase of system modeling (abstract system design); formal analysis, simulation and testing are not their priorities. Addressing properties 1 and 7 of CMP would lead to an extremely complex model.

The product-oriented (or state-oriented) paradigm addresses all conceptual properties of CMP except the 5th one—goal orientation—as this paradigm does not define the notion of goal. On the other hand, compared to goal-oriented formalisms, state-oriented modeling formalisms typically focus on concrete system design followed by validation and testing. They are supported by a plethora of techniques and tools for model simulation and formal analysis. Therefore, for modeling CMP, we adhere to the product-oriented paradigm.

3.4 Finite State Machines, Hierarchical State Machines and Statecharts

As explained above, we have chosen the product-oriented paradigm for modeling CMP. According to this paradigm, a state transition system (or state machine) represents our knowledge about the case and its evolution.

The choice of a concrete modeling formalism within the selected paradigm is related to the purpose of modeling (e.g., communication support, high-level design, simulation, formal validation and verification [18], diagnostics and improvement, recommendation and optimisation of process behaviour [8, 13]).

In this section we discuss a selection of existing formalisms based on state machines and focus on statecharts for CMP specification.

3.4.1 CMP Versus Complex Discrete-Event Systems

A CMP shares the following characteristics of complex reactive systems behavior defined in [19, 22]:

1. It continuously interacts with its environment. Its inputs and outputs are often asynchronous: they may occur or evolve unpredictably, at any time;
2. It must be able to respond to high-priority events (interrupts);
3. It has to operate and to react to inputs with respect to strict time regulations;
4. It has many possible operation scenarios, depending on its current mode of operation, current values of data as well as its past behavior;
5. It is very often based on interacting processes that operate in parallel.

As in a reactive system, the main challenge for the case manager is to identify the appropriate activities to perform in reaction to a given internal or external stimulus in a given situation (context).

State machines are a popular choice for specifying the behavior of reactive software systems. We will therefore consider them further.

3.4.2 Finite State Machines

A finite state machine (FSM) [38] specifies a machine that can be at one state at a time and can perform a state transition as a result of a triggering event (or a group of events guarded by a condition). It is defined by a (finite) set of states and a set of triggering events for each transition. To trigger a state transition, the execution of some activities and/or the observation of some contextual events can be required.

Traditional FSMs and their corresponding state-transition diagrams are very efficient for tackling small problems. However, the complexity of a FSM model tends to grow much faster than the complexity of the problem it represents. This makes the simulation or automated reasoning about the model extremely difficult. This phenomenon is called *the state explosion problem* [56].

3.4.3 Hierarchical State Machines and Statecharts

The state explosion problem can be overcome by the introduction of multiple hierarchical levels for states and transitions. Indeed, this hierarchy gives a possibility to reuse some common behaviors across many states and, thus, to reduce the model complexity. This idea is explored in the formalism of statecharts, invented by David Harel in the 1980s [19].

The statecharts formalism specifies a hierarchical state machine (HSM); it extends classical FSM by providing:

(i) depth—the possibility to model states at multiple hierarchical levels, with the notion of abstraction/refinement between levels;
(ii) orthogonality—the possibility to model concurrent or independent subma-chines within one state machine;
(iii) broadcast communication—the possibility to synchronize multiple concurrent submachines via events. Each internal (produced by the system) of external (produced by the environment) event is instantaneously broadcasted.

statecharts = FSM + Abstraction + Orthogonality + Broadcast-communication

Some state-oriented approaches (e.g., Petri Nets) associate a transition with the execution of one concrete activity (or a group of activities). On the contrary, with statecharts we associate a state transition with the occurrence of a triggering event (or combinations of events) allowing for a *deferred activity binding*. Thanks to the deferred binding, at design-time, the process scenario can be seen as a sequence of events; the concrete activities that will produce these events can be selected or invented in run-time. The process enactment can be seen as a dynamic selection of activities to produce some outcomes (events) that make the process progress towards its (desired) final state.

Visual notation. In the statechart notation, states are depicted with rectangular boxes with rounded corners. Figure 3.2 illustrates a high level diagram for our flood man-agement process example. The substate–superstate relation is depicted by boxes encapsulation. *Activation of Crisis Centers* and *EPSF* are exclusive substates of the *Flood Emergency* state: when in the *Flood Emergency* state, the case can be either in one or in the other of these substates. While entering the *Flood Emergency* state for the first time, the *Activation of Crisis Centers* substate is entered "by default"—this is depicted by the arrow with a black circle pointing at this substate.

Figure 3.3 shows a detailed diagram of the *EPSF* state from Fig. 3.2. The areas separated by the dashed lines represent the concurrent substates of their *EPSF* super-state: when in the *EPSF* state, the case is simultaneously in eight concurrent substates. Each of them can be seen as a separate statechart with its own state hierarchy. Thus,

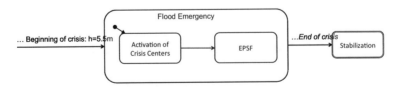

Fig. 3.2 High-level view of the Flood management process

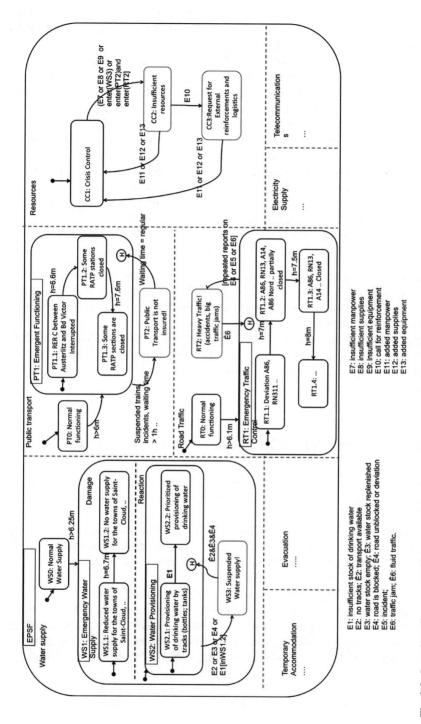

Fig. 3.3 Statechart diagram specifying the crisis management process once the EPSF is activated

E1: insufficient stock of drinking water
E2: no tracks; Ê2: transport available
E3: water stock empty; Ê3: water stock replenished
E4: road is blocked; Ê4: road unblocked or deviation
E5: incident;
E6: traffic jam; Ê6: fluid traffic.

E7: insufficient manpower
E8: insufficient supplies
E9: insufficient equipment
E10: call for reinforcement
E11: added manpower
E12: added supplies
E13: added equipment

the introduction of concurrent substates is a convenient mechanism to specify logically different areas of the case management (Public Transport management, Water Supply management, Road Traffic management etc.).

The set of active states of all concurrent substates is called the *active configuration* of a statechart. It replaces the term of "current state" in conventional (flat) FSM.

The transition that terminates with a circle with "H" stands for "entering the state by history". The transition from *PT1* to *PT2* in Fig. 3.3, for example, specifies that once the case recovers from the *PT2: Public Transport is Not insured* state and re-enters the *PT1:Emergent Functioning* state—the last active configuration of the latter is selected (and not the default one).

The transitions between states in statecharts are depicted by arrows labeled with expressions that specify the triggering events and (optionally) the actions that are carried out while the transition is triggered. In our example, the triggering events mostly represent external and internal events.

More details on the statecharts notation can be found in [23]. The semantics of statecharts for CMP will be presented in more details in Sect. 3.5.

Execution of statecharts specifications. The operational semantics of statecharts was originally implemented in the STATEMATE system and described in [21, 31]. The statecharts formalism was also adopted by the UML community in the form of UML statemachine diagrams [46].

Rhapsody [20] (now IBM Rational Rhapsody) and open source YAKINDU Statechart Tools (SCT) are examples of statecharts modeling environments, where the statecharts specifications can be created and executed in an intuitive and interactive way.

3.5   Statecharts Semantics for Case Management Processes

As explained above, we adopt the formalism of statecharts for the specification of case management processes (Sect. 3.5.1). We also propose some extensions of statecharts in Sect. 3.5.2.

We create the statecharts specification for the flood management process based on the description provided by the Emergency Plan Specialized on Floods (EPSF) and on some practical knowledge about resource management during floods. The resulting diagrams are shown in Figs. 3.2 and 3.3.

We start with a high-level view of the process described by two states—*Flood Emergency* and *Stabilization*—and transitions between them (Fig. 3.2). The *Flood Emergency* state is entered when the water level at Austerlitz Bridge raises above 5.5 m. It contains two substates: *Activation of Crisis Centers* and *EPSF*. The transition to *Stabilization* state is triggered once specific conditions identified as "*end of crisis*" are met.

The diagram in Fig. 3.3 specifies the main areas of crisis management as concurrent substates of the *EPSF* state. For the purpose of this work, we show only a few of these substates in detail: *Water supply*, *Public transport*, *Road traffic* and *Resources*. This model can be refined providing further details on the crisis management scenarios and operation procedures.

3.5.1 Statecharts Semantics for CMP Specification

Below, we explain how the following concepts defined by the statecharts formalism [23] can be adopted for the specification of CMP:

- State, state hierarchy and state decomposition;
- Abstraction and refinement;
- AND, OR and basic states;
- Entering a state by default and by history;
- State configuration;
- Internal, external and triggering events;
- Activity;
- Broadcast communication;
- Inter-level transition.

State, state hierarchy and state decomposition. A CMP state can be seen as a specific situation in the case management process that requires reaction.

On the abstract level, states can be compared to business milestones. The definition of the right set of states for the process is subjective: it reflects our current understanding of the process and evolves over time. In this work, the states of the flood crisis management process are characterized by the level of water h. These states represent the critical points for different management areas defined by EPSF (Sect. 3.2).

While *being in a given state*, some work has to be done in order to maintain this state or in order to leave this state and enter another state. Note that statecharts do not specify how exactly this work will be performed or which activities will be executed and in which order. Another means for modeling activities is needed: statecharts, for example, can be complemented with activity charts [23]. In this paper, we do not discuss activity modeling in detail.

In Fig. 3.3, three states *RT0*, *RT1* and *RT2* specify the main phases of the road traffic control after the emergency plan (EPSF) is triggered.

- *RT0: Normal functioning* is the default state upon triggering the EPSF. The water level of 5.5 m does not disrupt the road infrastructure of the region and normal functioning is maintained.
- *RT1: Emergency traffic control*—this state is attained at 6.1 m; here the flood is affecting the road traffic. Specific measures must be continuously taken in this state in order to maintain road safety.

- *RT2: Heavy Traffic!*—this state is reached when the road traffic degrades (due to accidents, traffic jams) to the point where the crisis management itself becomes compromised (e.g., the rescue teams cannot arrive to the endangered areas, etc.).

A state s consists of a (possibly empty) *hierarchy* of substates, representing (possibly concurrent) state machines. These substates provide details about their parent state (or superstate).

In Fig. 3.3, four different substates (from *RT1.1* to *RT1.4*) are defined based on the flood severity: upon entering each of these states, the city executes some scenario: deploying equipment, marking deviations, blocking roads, informing drivers, etc. Each substate belongs to one superstate (its surrounding state) that is also its nearest ancestor in the state hierarchy. We call the relations between the superstate and its substates abstraction/refinement relations.

Abstraction and refinement. State abstraction consists in clustering states into a superstate according to some similarity criteria. This mechanism allows one to describe the problem at multiple abstraction levels, hiding or introducing details when necessary. Refinement is the opposite of abstraction, it consists in decomposing a state into substates according to some discrimination criteria.

More formally, *refinement* is a XOR decomposition of a state, where being in a superstate means being in exactly one of its (exclusive) component substates.

One substate can be marked as default so that this state is visited each time its parent state is entered.

Public Transport Emergent Functioning state (PT1) in Fig. 3.3 is specified with three exclusive substates corresponding to three different management scenarios that are activated based on the water level h. *PT1.1* is the default scenario.

From a visualization standpoint, clustering states allows for a very economical representation. It avoids duplicating transitions and the model logical structure appears clearly.

The AND decomposition results in the specification of orthogonal (or concurrent) components of the parent state. The AND decomposition models the situation when being in the state means being in one of the combinations of its components. All possible combinations make an orthogonal product.

AND, OR and basic states. The statecharts formalism defines three types of states: AND, OR and basic states. The AND-state is a state that contains two or more orthogonal substates; the OR-state is a state that contains one or more exclusive substates. A state is basic if it does not have any substates.

Consider the *Water Supply* in Fig. 3.3: *WS1:Emergency water supply* is an AND-state that contains two concurrent substates *Damage* and *Reaction*. These substates model the damage due to the flood and the reaction to it, i.e., emergency water provisioning. Once *WS1* is entered both of its concurrent substates are activated.

The *Damage* state is an OR-state; it contains two exclusive substates that specify its details: *WS1.1*, where the water supply of some towns is reduced; *WS1.2*, where the water supply is totally disrupted.

The *Reaction* state is an OR-state; it contains two exclusive substates: *WS2: Water provisioning* and *WS3: Suspended Water supply!*

Once *WS1* is entered, *WS1.1, WS2* and its substate *WS2.1* are entered by default.

Emergent water provisioning (WS2) defines specific measures to provide areas with drinking water: normal provisioning (WS2.1) and prioritized provisioning (WS2.2) in case of limited stock of drinking water (event *E1*). The state *WS3: Suspended Water Supply!* has no substates—it is a basic state. It refers to the situation when the emergent water provisioning can no longer be guaranteed. This, for instance, can result from severe road conditions or insufficient stock of bottled drinking water while no other supply is available (i.e., when the case is in *WS1.2* state). This is indicated by the transition label: *E2 or E3 or E4 or E1 [in WS1.2]]*.

Entering a state by default and by history. The default indicator is used to identify which substate will be visited when its parent state is entered. Alternatively, in many cases it can be useful to enter the superstate by history, i.e., to enter its most recently visited substates (or configuration of substates). The examples include the transition from *WS3* back to *WS2* in Fig. 3.3: once the problem is solved, the emergency water supply is restored at its latest visited substate (which is not necessarily *WS2.1*).

State configuration. Compared to conventional (flat) FSM, in hierarchical state machines depicted by statecharts, multiple states can be activated at the same time. Statecharts define the term *configuration* (of a state or of a system):

The *active configuration* of a state s is the set of basic substates of s that are activated at the current moment. Intuitively, active configuration replaces the conventional term of current state defined in FSM.

For the state WS1 in Fig. 3.3, consider that $h = 7\,\text{m}$ and the water provisioning functions normally; this would correspond to the following active configuration of $WS1$: $cf(WS1) = WS1.2, WS2.1$

The sequence of active configurations resulting from the execution of the statechart specification represents a *trace* of the CMP.

Internal, external and triggering events. *Internal events* are produced by the system (Fig. 3.1); they are the results of carried out activities.

In Fig. 3.3, event *E1* specifies an *insufficient stock of drinking water*. It is an internal event that can result from the water distribution activity or can be generated by some other activity like stock verification.

External events are produced by the environment (context) of the case. The case context consists of various objects that influence the case and affect its handling (Fig. 3.1). Water level, weather forecast, current situation on the roads, incident reports are examples of contextual parameters *sensed* by a system during a flood.

In Fig. 3.3, *WS0–WS1* or *WS1.1–WS1.2* state transitions are taken if a certain value of h (water level) is reached or exceeded. We consider that the change in the water level is an external (contextual) event.

The *triggering event e[c]* (interpreted as *e occurs and c holds*) of a transition t is an event that must occur in order for t to take place. Here $e \in E$ is the event (or a logical combination of events) that triggers the transition; $c \in C$ is a condition that needs to be true for the transition to be taken when e occurs.

In Fig. 3.3, the triggering event for the transition from *WS2* to *WS3* is described by an expression: *E2 or E4 or E1[in(WS1.2)]*. This transition is taken if no more tracks for transporting bottled water are available or if the road to the concerned area is blocked or if the stock of drinking water on place is insufficient while some towns no longer have regular water supply. The operational information about the resources or traffic conditions corresponds to contextual or internal events. Condition *in(WS1.2)* specifies that *WS1.2* substate is active.

To specify some work to be done, statecharts use the concept of activity.

Activity. The statecharts formalism defines *state-dependent activities* that are linked directly to a state *s* and can be carried out *throughout* or *within s*. In the first case, an activity starts when entering the state *s* and terminates when leaving it. In the second case, an activity starts when entering *s*; when exiting *s*, if the activity it is not terminated yet, it is stopped by the system.

This is a valid interpretation for a case management process too: in our example, upon entering the state *WS2.1* (in Fig. 3.3) an activity for water provisioning must be started and must continue within this state. Upon entering *PT1*, activities for closing the concerned stations must be carried out *throughout* this state.

Relations between activities and states defined by statecharts can be characterized as mandatory: if activity *A* is linked to state *s* by a *throughout* or a *within* relation it must be carried out at this state. Therefore, each state *s* can be associated with a (possibly empty) set of mandatory activities.

Broadcast communication. Broadcasting allows for communication and synchronization between concurrent sub-machines. According to statecharts, both internal and external events are *broadcasted*, meaning that one single event can trigger transitions at multiple orthogonal substates.

Broadcast communication allows for coordination between different management areas in CMP. For example, *Road Traffic* and *Resources* substates of *EPSF* can both react to an (external) event reporting on the hard traffic in a particular road section. Along those lines, blocked roads or insufficient water supply (internal events) may trigger evacuation of people and facilities from the concerned area.

Inter-level transitions. The transitions that cross state boundaries are called inter-level transitions in statecharts. Their purpose is to model the interruptions or the situations when the process has to react, no matter its state or the activity it performs. For example, if the *End of crisis* event occurs (Fig. 3.3), no matter what the configuration of *EPSF* substates is and what activities are executed in all its areas, they will be terminated and the new state (presumably *Stabilization*) will be entered by the process.

3.5.2 Adaptation and Extension of the Statecharts Formalism for CMP Specification

Below, we discuss the specific features of CMP that cannot be captured by the original statecharts concepts and we therefore propose adaptations and extensions to the statecharts formalism.

What kinds of extensions are needed? Why? Despite the similarities identified in Sect. 3.4, there exists a number of characteristics that makes a CMP significantly different from a conventional reactive system:

1. A CMP has a goal. In reaction to given stimuli in a given situation, the case manager searches for scenarios that could steer the case towards its goal. Thus, compared to a reactive system where the next state (or active configuration) is defined by its current state and a given situation, the next state in CMP is also defined by the process goal.
2. CMP is a knowledge-intensive process where decisions (e.g., scenario planning, task assignment) are typically made by the case manager. As a consequence:
3. CMSS can be compared to business-intelligence systems (BI) rather than auto-mated control systems: for the latter, once the preconditions are satisfied for an activity a, a is *automatically executed*. For CMSS, once the preconditions of a are satisfied, a is *enabled for execution* and (unless explicitly stated as *mandatory*) the case manager decides weather it will be carried out or not.
4. Whereas some events are relevant only immediately after they occur (e.g., button pressed), other events, once they occur, remain relevant or valid for some period of time or for the whole execution of a CMP (e.g., document received; permission granted).

In order to faithfully represent the complexity of a CMP, we propose to extend the statecharts formalism with the following concepts:

1. Final configuration;
2. Path, path selection and path reinforcement;
3. Relevance and validity interval for events;
4. Mandatory versus optional activities.

We briefly describe these concepts in the remainder of this section.

Final configuration. Similarly to [51], we express the process *goal* in terms of "target state" (or configuration) that the state transition system has to reach or to maintain; strategies (possible scenarios) for achieving this goal can be seen as sequences of states to visit (or state transitions to fire) between the "current" and the "target" states.

By analogy with the active configuration defined by statecharts, we define a final configuration for CMP:

The *final configuration* of a state s is the set of basic substates of s that we want to enter and/or maintain upon the CMP termination.

In our example, the goal of the process is to support the areas affected by the flood and to protect the population from the flood consequences. For the statechart in Fig. 3.3, any configuration where the critical states *WS3, PT2, RT2, CC2, CC3* are not active can be considered as a final configuration (i.e., it should be maintained until the crisis is over).

Path, path selection and path reinforcement. We define a *path* in statecharts as any sequence of active configurations that terminates with the final configuration.

In our example, if one of the concurrent submachines has entered a critical state, the path is the sequence of configurations that would lead this submachine back to one of its non-critical states. For example, if the state *CC2: Insufficient resources* of the *Resources* substate is active, there are two paths for this submachine to bring it back to CC1: $CC2 \longrightarrow CC1$ or $CC2 \longrightarrow CC3 \longrightarrow CC1$.

In any given active configuration, a path towards the final configuration can be calculated. The optimal path can be selected using some criterion (e.g., the cheapest path, the shortest path, the path with the highest probability to be realized).

Consider the optimal path p from the current active configuration to the final configuration. *Enforcing path p* means executing some activities in order to enable and then to take some state transition t that will lead to the next active configuration in p.

Consider that the state *CC2: Insufficient resources* of the *Resources* submachine is active and that the path $CC2 \longrightarrow CC1$ is the optimal path. Enforcing this path means here enabling the transition from *CC2* to *CC1*. This transition will be taken if at least one of the events *E11* (added manpower), *E12* (added supplies) or *E13* (added equipment) occurs. We can enforce the path by mobilizing volunteers or relocating supplies, manpower or equipment, considering that these activities can generate E11, E12 or E13 as a result.

Relevance and validity interval for events. To take some transition t, the execution of process activities and/or observation of contextual events can be required.

Most process formalisms including statecharts define a triggering event as a single event, or a group of events that occur simultaneously and instantly trigger a state transition. In particular, statecharts specify that *events are only available in the step directly succeeding their generation* [19]. We call such events **instantaneous** events and distinguish them from **continuous** events that, once observed, remain valid and can be reacted upon asynchronously, during multiple steps.

We define the *validity interval tv* for event e as a period of time between the moment when this event is first observed and the moment when it becomes irrelevant for the process.

To define the validity interval for an event e, we associate it with the time the system resides in some state or with the occurrence of another event $\hat{e}$ that cancels e:

If the validity interval tv of event e is some state s: $tv(e) = s$, this means that, after being sensed for the first time in s (or one of its substates), e will be valid until the system leaves s. The higher the state in the state hierarchy, the longer the validity interval.

If the validity interval tv of event e is some event $\hat{e}$: $tv(e) = \hat{e}$, this means that, after being sensed, e will be valid until $\hat{e}$ occurs.

For example, the *approval received* event for starting the evacuation procedure (not shown in the statechart in Fig. 3.3) is valid as long as the *Flood Emergency* is active; *E6* (traffic jam) event is valid until *Ê6* (fluid traffic) event is received.

If no validity interval is specified for an event, this event is an *instantaneous* event.

We extend the definition of triggering event as follows: *the triggering event e[c]* of a transition *t* is an event that must occur for *t* to take place. Here $e \in E$ is a combination of events and/or absence of events observed during some period of time (validity interval) that triggers the transition; $c \in C$ is a condition that needs to be true in order the transition to be taken when *e* occurs.

This definition allows us to take into account not only immediate events but also relevant events observed in the past (email received, approval obtained, etc.).

Mandatory versus optional activities. In the statecharts formalism, activities can be considered as *state-dependent* [56] (i.e., each state s is associated with a list of activities to be carried out in this state). These activities are also *mandatory*: they are automatically executed once their preconditions are met.

To relax the coupling between *a situation* and *a reaction* and to allow for more flexibility in process execution, we propose to define *state-independent* activities for statecharts—activities, that can be executed in any configuration if their preconditions are met (unless explicitly stated otherwise).

Thus, each state *s* can be associated with two (possibly empty) sets of activities: the set of *mandatory* activities that must be carried out in *s* and are state-dependent and the set of optional or *enabled* activities that are defined dynamically, based on the statechart status. Optional activities can be executed *within* or *throughout* the state in order to ensure the right progression of the case towards its goal.

The specification of activities is beyond the scope of this chapter.

3.6   Perspectives and Roadmap for Future Research

The benefits of the proposed formalism are numerous for CMP: executable statecharts specifications allow for interactive design, simulation-based testing and simulation-based recommendations. In the future, these features could be integrated as a part of CMP-supporting PAIS in order to provide intelligent decision-support functionalities for case managers. To conclude this chapter we discuss these perspectives and outline the directions for future work.

3.6.1   Design and Simulation-Based Testing

Interactive design of CMP. A statecharts specification can be created based on some a priori knowledge about the CMP (e.g., norms, regulations, best practices, etc.) (Fig. 3.4a). Thanks to the concept of hierarchical state, this model can be extended

Fig. 3.4 Incremental design of CMP

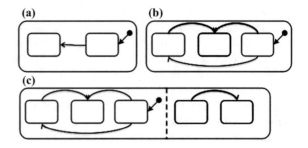

and refined by integrating the experience of the case manager: new states and state transitions can be specified reflecting new situations and the way to deal with them (Fig. 3.4b); concurrent substates can be added in order to increase the scope of the process (Fig. 3.4c). IBM Rational Rhapsody[10]) and open source YAKINDU State-chart Tools (SCT)[11] are examples of statecharts modeling environments, where the statecharts specifications can be created and executed in an intuitive and interactive way. However, creating a detailed statecharts specification for a real-life CMP is a challenging task, as explained below.

Using clustering techniques for statecharts improvement. Although statecharts have been designed to represent states in a hierarchical way, this formalism does not specify how states should be organized into abstraction levels. In most cases this clustering of states is performed manually by a process designer.

Clustering algorithms gather entities (e.g., the states of a state machine) into clusters according to some similarity criteria that can include complex sets of parameters. Formal Concept Analysis (FCA) [17] is a well-known clustering technique that is successfully used in many areas including knowledge discovery, representation and sharing [39]. In FCA, the obtained clusters are organised into a lattice using generalisation and specialisation relationships, which could be used to identify state hierarchy in statecharts. A significant advantage of FCA is that the resulting clusters may overlap, whereas many traditional clustering techniques build partitions.

FCA can therefore be used for clustering states and helping to define the hierarchical structure within a statecharts specification. Various attributes may be chosen to describe states: pre-conditions, post-conditions, contextual parameters, and any combination of them. As a result, each FCA concept (cluster) is explicitly labelled by the set of attributes that characterize the objects of the cluster. This can be considered as a starting point for further model analysis and improvement: detection of "missing" states or state transitions, identification of "similar" states or activities etc.

Simulation of CMP specifications. Statecharts combine an intuitive and concise visual notation with precise semantics. Thanks to these semantics, the statecharts

---

[10]http://www-03.ibm.com/software/products/en/ratirhapfami.

[11]http://statecharts.org/index.html.

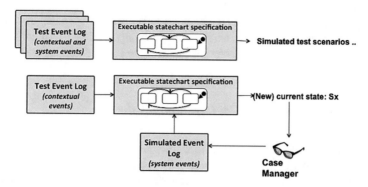

Fig. 3.5 Simulation-based testing of CMP scenarios

specifications can already be simulated at the early design stages, providing an instant visual feedback. At the later design stages, they can serve as a basis for simulation-based testing as explained below.

A statecharts specification can be executed with a test event log (i.e., a pre-recorded sequence of events defining some flood development scenario) allowing for the simulation and testing of various handling scenarios. Two simulation modes can be defined:

- A fully automated mode (Fig. 3.5a), where the statecharts specification is executed with a test event log that includes both contextual events (e.g., raise of water level, traffic jam) and system events (e.g., successful deployment of equipment, empty stock of drinking water). The events from the event log are processed by a statecharts simulation environment[12] triggering the state transitions. The simulation result is a sequence of visited states.
- In an interactive mode (Fig. 3.5b), the test event log contains only contextual (external) events and emulates the environment. A case manager reacts to external events by executing enabled activities (e.g., deploying equipment, making task assignments)—these activities represent the steps of case handling scenarios.

Similarly to computer simulator games, the interactive statecharts simulation is an iterative process, where the case response is simulated after each step taken by the case manager: pre-recorded external events and internal events resulting from the case manager's decisions trigger state transitions in statecharts specification and once the (new) current state **Sx** is entered a new step starts. The simulation result is the sequence of visited states, executed activities and received events.

In the case of a crisis management process, multiple scenarios can be "played" automatically or interactively and used as a basis for trainings, drills and improvement of formal operation procedures (e.g., procedures described by EPSF).

---

[12]Development of modeling and simulation environment for CMP will be addressed in our future work.

Conversely, possible case development scenarios can be calculated as sequences of events acceptable by the state machine representing the CMP. This could help to analyse the process and reveal scenarios that were not considered before.

3.6.2 Simulation-Based Recommendations

Gartner's Hype Cycle for Emerging Technologies report provides a cross-industry perspective on technologies and trends, with an assessment of their maturity, adoption and business benefit. According to the reports from 2013 and 2014,[13] Predictive Analytics technologies have already reached their plateau of productivity and are currently becoming the mainstream technology, whereas Complex-Event Processing (CEP), Big Data and Content Analytics are currently rolling down from their peak of inflated expectations and will reach their maturity (the plateau) in 5 to 10 years. This makes run-time situation analysis and recommendations for case managers the next challenge for the CMP-supporting PAIS.

Some recommendation systems supporting process modeling and process management are presented in the literature [29, 50]. Process mining is a widely recognised technique for predicting a best process scenario based on the analysis of past execution logs [4]. The approach reported in [8] uses constraint-based programming to generate recommendations on process execution strategies.

An example of CMP solution integrating intelligent support for the case manager is reported in [26]. Here the authors introduce the concept of User-Trained Agent (UTA), which recommends the best next actions based on the actions taken by the case managers in previous similar situations. The proposed recommendation technique is based on pattern recognition and is integrated as a part of ISIS Papyrus platform.

Whereas all the approaches for recommendations mentioned above are based on "past experience" or process logs, we propose an alternative technique that is based on *the execution of a CMP specification in the simulated process environment*:

In our vision, the "a priory" statecharts specification of a CMP can be analysed using graph theory algorithms. The objective of the analysis is to search and optimize a path from some current state of the statecharts model to its target state, representing the goal of the process. As a result, the "best next state to visit", "best next transition to fire" and, consequently, "possible activities to execute" are recommended to the case manager. The main advantage of this analysis is that recommendations can be provided based on:

- our current knowledge about the process represented by its executable statecharts specification and
- our current knowledge about the process environment, represented by a real-time event buffer (or event log).

---

[13]Gartner, http://www.gartner.com/newsroom/id/2575515, http://www.gartner.com/newsroom/id/2819918.

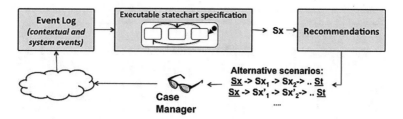

Fig. 3.6 Run-time recommendations on the CMP activity planning

No "past experience" represented by a log of the past process executions is required. This makes "cold starts" possible.

A statecharts specification could be initialized and then executed using a *real-time event log* (i.e., where both contextual (external) and system (internal) events occur in real time). Given a current state **Sx** of the statecharts and the desired (target) state **St**, possible case management scenarios can be calculated as alternative *paths* from **Sx** to **St** on the statecharts diagram. Each scenario can be seen as a sequence of "correct" state transitions resulting from the execution of corresponding activities (Fig. 3.6).

The alternative scenarios and activities that need to be executed in order to reinforce these scenarios could then be recommended to the case manager.

The integration of CMP executable specifications and analysis tools as a part of CMP-supporting PAIS could provide an intelligent support for case managers, as explained below.

3.6.3 Enhancing the CMP-supporting PAIS with Recommendations for Agile Activity Planning

Figure 3.7 illustrates our *vision* of the intelligent CMSS introduced in our earlier works [48, 49]. We describe below the main components of this system: Dynamic context manager (DCM), Navigation manager (NM), Activity/Resource repository, Log and History.

The role of *the Dynamic context manager* is to select, measure and monitor relevant contextual variables of the CMP. Internal and external events are collected and stored in the *Event log*. The *Activity/Resource repository* stores the definitions of activities that can be performed during the case handling and resources that can be used. The *Log and History* component registers the ongoing process scenario (i.e., the sequence of executed activities, received events and visited states).

The *Navigation Manager* is the "heart" of the system, it provides intelligent support for the case manager by recommending the best scenario(s) for handling the case. It contains the *Executable statecharts specification* of the CMP and the *Recommendation* component that uses graph theory, process mining and clustering algorithms in order to provide recommendations for the case manager.

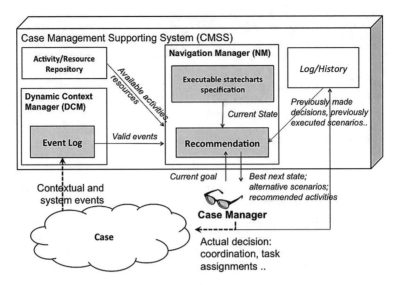

Fig. 3.7 Intelligent CMSS. The Navigation Manager provides recommendations about the best scenario based on the current state of the process and the list of valid events

The *Executable statecharts specification* of CMP models the *behavior* of the SUD and its Environment. It can be executed with the collected real-time events. Possible case management scenarios are described by the sequences of states of the statecharts model that lead from the current state to some target state that represents the CMP objective.

The *Recommendation* component can provide the case manager with an insight about how the situation might develop and about the possible strategies (paths in statecharts, activities, groups of activities to carry out) to bring the situation under control. The recommendation mechanism uses the Activity/Resource repository to define the list of activities enabled at a given situation identified with the current configuration (state) of the statecharts model. Since activities are independent from states, new activities can be added to the Activity/Resource repository at run time and further used by the Recommendation component without needing to change the model.

The intelligent CMSS sketched above will be grounded on the statecharts specifications enabling incremental interactive process design, simulation-based testing and recommendations. According to the statecharts formalism, a case management process is represented by a hierarchical state machine, where the process scenario can be seen as *a dynamic choice of activities with an objective to trigger a "good" state transition that would move the case from its "current state" towards its "target state" representing the case management goal.*

The development of a prototype of the intelligent CMSS is our future work.

References

1. van der Aalst, W.: The application of petri nets to workflow management. J. Circ. Syst. Comput. **8**(01), 21–66 (1998)
2. van der Aalst, W.: Process-aware information systems: lessons to be learned from process mining. In: Jensen, K., van der Aalst, W. (eds.) Transactions on Petri Nets and Other Models of Concurrency II. LNCS, vol. 5460, pp. 1–26. Springer (2009)
3. van der Aalst, W., Ter Hofstede, A.H.: Yawl: yet another workflow language. Inf. Syst. **30**(4), 245–275 (2005)
4. van der Aalst, W., Weijters, A.: Process mining: a research agenda. Comput. Ind. **53**(3), 231–244 (2004)
5. van der Aalst, W., Pesic, M., Schonenberg, H.: Declarative workflows: balancing between flexibility and support. Comput. Sci. Res. Dev. **23**(2), 99–113 (2009)
6. van der Aalst, W., Weske, M., Grünbauer, D.: Case handling: a new paradigm for business process support. Data Knowl. Eng. **53**(2), 129–162 (2005)
7. Bajec, M., Krisper, M.: A methodology and tool support for managing business rules in organisations. Inf. Syst. **30**(6), 423–443 (2005)
8. Barba, I., Weber, B., Del Valle, C.: Supporting the optimized execution of business processes through recommendations. In: Daniel, F., Barkaoui, K., Dustdar, S. (eds.) Business Process Management Workshops. LNBIP, vol. 99, pp. 135–140. Springer, Berlin (2012)
9. Bider, I.: Towards a non-workflow theory of business processes. In: La Rosa, M., Soffer, P. (eds.) Business Process Management Workshops. LNBIP, vol. 132, pp. 1–2. Springer, Berlin (2013)
10. Bubenko, J., Rolland, C., Loucopoulos, P., DeAntonellis, V.: Facilitating fuzzy to formal requirements modelling. In: Proceedings of the First International Conference on Requirements Engineering, 1994, pp. 154–157. IEEE (1994)
11. Cauvet, C.: Modélisation des processus d'ingénierie des systèmes d'information. Encyclopédie de l'Informatique et des Systèmes d'Information, pp. 1412–1425 (2006)
12. Davenport, T.: Thinking for a Living: How to Get Better Performances and Results from Knowledge Workers. Harvard Business Press (2005)
13. Dijkman, R., Dumas, M., Garca-Bauelos, L.: Graph matching algorithms for business process model similarity search. In: Dayal, U., Eder, J., Koehler, J., Reijers, H. (eds.) Business Process Management. LNCS, vol. 5701, pp. 48–63. Springer, Berlin (2009)
14. Dijkman, R.M., Dumas, M., Ouyang, C.: Semantics and analysis of business process models in BPMN. Inf. Softw. Technol. **50**(12), 1281–1294 (2008)
15. Dowson, M.: Iteration in the software process; review of the 3rd international software process workshop. In: Proceedings of the 9th International Conference on Software Engineering, ICSE '87, pp. 36–41. IEEE Computer Society Press, Los Alamitos, CA, USA (1987)
16. Dumas, M., van der Aalst, W.M., Ter Hofstede, A.H.: Process-Aware Information Systems: Bridging People and Software Through Process Technology. Wiley (2005)
17. Ganter, B., Wille, R., Wille, R.: Formal Concept Analysis, vol. 284. Springer, Berlin (1999)
18. Groefsema, H., Bucur, D.: A survey of formal business process verification: from soundness to variability. In: Proceedings of International Symposium on Business Modeling and Software Design (BMSD) (2013)
19. Harel, D.: Statecharts: a visual formalism for complex systems. Sci. Comput. Program. **8**(3), 231–274 (1987)
20. Harel, D., Gery, E.: Executable object modeling with statecharts. In: Proceedings of the 18th International Conference on Software Engineering, ICSE '96, pp. 246–257. IEEE Computer Society, Washington, DC, USA (1996)
21. Harel, D., Naamad, A.: The statemate semantics of statecharts. ACM Trans. Softw. Eng. Methodol. **5**(4), 293–333 (1996)
22. Harel, D., Pnueli, A.: On the development of reactive systems. Springer (1985)
23. Harel, D., Politi, M.: Modeling Reactive Systems with Statecharts: The STATEMATE Approach. McGraw-Hill, Inc. (1998)

24. Hull, R., Damaggio, E., De Masellis, R., Fournier, F., Gupta, M., Heath, III, F.T., Hobson, S., Linehan, M., Maradugu, S., Nigam, A., Sukaviriya, P.N., Vaculin, R.: Business artifacts with guard-stage-milestone lifecycles: managing artifact interactions with conditions and events. In: Proceedings of the 5th ACM International Conference on Distributed Event-Based System, DEBS '11, pp. 51–62. ACM, New York, NY, USA (2011)
25. Kemsley, S.: The changing nature of work: from structured to unstructured, from controlled to social. In: Rinderle-Ma, S., Toumani, F., Wolf, K. (eds.) Business Process Management. LNCS, vol. 6896, pp. 2–2. Springer, Berlin (2011)
26. Kim, T.T.T., Ruhsam, C., Pucher, M.J., Kobler, M., Mendling, J.: Towards a pattern recognition approach for transferring knowledge in ACM. In: Enterprise Distributed Object Computing Conference Workshops and Demonstrations (EDOCW), pp. 134–138 (2014)
27. Kirsch-Pinheiro, M., Rychkova, I.: Dynamic context modeling for agile case management. In: Demey, Y., Panetto, H. (eds.) On the Move to Meaningful Internet Systems: OTM 2013 Workshops. LNCS, vol. 8186, pp. 144–154. Springer, Berlin (2013)
28. Koehler, J., Tirenni, G., Kumaran, S.: From business process model to consistent implementation: a case for formal verification methods. In: Enterprise Distributed Object Computing Conference, 2002, EDOC '02. Proceedings, pp. 96–106 (2002)
29. Koschmider, A., Oberweis, A.: Designing business processes with a recommendation-based editor. In: Brocke, J., Rosemann, M. (eds.) Handbook on Business Process Management 1. International Handbooks on Information Systems, pp. 299–312. Springer, Berlin (2010)
30. van Lamsweerde, A.: Goal-oriented requirements engineering: a guided tour. In: 2001. Proceedings. Fifth IEEE International Symposium on Requirements Engineering, pp. 249–262 (2001)
31. Mikk, E., Lakhnech, Y., Petersohn, C., Siegel, M.: On formal semantics of statecharts as supported by statemate. In: Workshop, Ilkleym, vol. 14, p. 15 (1997)
32. Murata, T.: Petri nets: properties, analysis and applications. Proc. IEEE **77**(4), 541–580 (1989)
33. Nurcan, S., Edme, M.H.: Intention-driven modeling for flexible workflow applications. Softw. Process: Improv. Pract. **10**(4), 363–377 (2005)
34. OMG: Case management process modeling (CMPM) request for proposal. http://www.omg.org/cgi-bin/doc?bmi/09-09-23 (2009)
35. OMG: Business process model and notation (BPMN). http://www.omg.org/spec (2011)
36. OMG: Case management model and notation. http://www.omg.org/spec/CMMN/1.0/PDF/ (2014). document number formal/2014-05-05
37. Pesic, M., Schonenberg, H., van der Aalst, W.: Declare: full support for loosely-structured processes. In: Enterprise Distributed Object Computing Conference, 2007, EDOC 2007. 11th IEEE International, pp. 287–287. IEEE (2007)
38. Plotkin, G.: A structural approach to operational semantics (1981)
39. Poelmans, J., Elzinga, P., Viaene, S., Dedene, G.: Formal concept analysis in knowledge discovery: a survey. In: Conceptual Structures: From Information to Intelligence, pp. 139–153. Springer (2010)
40. Pohl, K., Weidenhaupt, K.: A contextual approach for process-integrated tools. In: Jazayeri, M., Schauer, H. (eds.) Software Engineering ESEC/FSE'97. LNCS, vol. 1301, pp. 176–192. Springer, Berlin (1997)
41. Pucher, M.: The elements of adaptive case management. In: Mastering the Unpredictable, pp. 89–134 (2010)
42. Reijers, H.A., Limam, S., Van Der Aalst, W.: Product-based workflow design. J. Manag. Inf. Syst. **20**(1), 229–262 (2003)
43. Rolland, C., Prakash, N., Benjamen, A.: A multi-model view of process modelling. Requir. Eng. **4**(4), 169–187 (1999)
44. Rolland, C., Souveyet, C., Moreno, M.: An approach for defining ways-of-working. Information Systems **20**(4), 337–359 (1995)
45. Rosemann, M., van der Aalst, W.: A configurable reference modelling language. Inf. Syst. **32**(1), 1–23 (2007)

46. Rumbaugh, J., Jacobson, I., Booch, G.: Unified Modeling Language Reference Manual, 2nd edn. Pearson Higher Education (2004)
47. Rychkova, I.: Exploring the alloy operational semantics for case management process modeling. In: 2013 IEEE Seventh International Conference on Research Challenges in Information Science (RCIS), pp. 1–12 (2013)
48. Rychkova, I., Kirsch-Pinheiro, M., Le Grand, B.: Context-aware agile business process engine: foundations and architecture. In: Nurcan, S., Proper, H., Soffer, P., Krogstie, J., Schmidt, R., Halpin, T., Bider, I. (eds.) Enterprise, Business-Process and Information Systems Modeling. Lecture Notes in Business Information Processing, vol. 147, pp. 32–47. Springer, Berlin Heidelberg (2013)
49. Rychkova, I., Le Grand, B., Kirsch-Pinheiro, M.: Adaptive case management: supporting knowledge intensive processes with it systems. In: Fischer, L. (ed.) Empowering Knowledge Workers. BPM and Workflow Handbook Series. Future Strategies Inc. (2013)
50. Schonenberg, H., Weber, B., van Dongen, B., van der Aalst, W.: Supporting flexible processes through recommendations based on history. In: Dumas, M., Reichert, M., Shan, M.C. (eds.) Business Process Management. LNCS, vol. 5240, pp. 51–66. Springer, Berlin (2008)
51. Soffer, P., Yehezkel, T.: A state-based context-aware declarative process model. In: Enterprise, Business-Process and Information Systems Modeling, pp. 148–162. Springer (2011)
52. Swenson, K.: Mastering The Unpredictable: How Adaptive Case Management Will Revolutionize the Way That Knowledge Workers Get Things Do. Meghan-Kiffer Press (2010)
53. Swenson, K., Palmer, N., Manuel, A., Carlsen, S.: Empowering Knowledge Workers. BPM and Workflow Handbook Series. Future Strategies Inc. (2013)
54. Swenson, K., Palmer, N., Pucher, M., Manuel, A., Webster, C.: How Knowledge Workers Get Things Done. Future Strategies Inc. (2012)
55. Taylor, F.W.: The principles of scientific management. Harper (1914)
56. Wagner, F., Schmuki, R., Wagner, T., Wolstenholme, P.: Modeling software with finite state machines: a practical approach. CRC Press (2006)
57. Weske, M.: Business Process Management: Concepts, Languages, 2nd edn. Architectures. Springer, Berlin (2012)
58. Yu, E.S.: Towards modelling and reasoning support for early-phase requirements engineering. In: Requirements Engineering, 1997., Proceedings of the Third IEEE International Symposium on. pp. 226–235. IEEE (1997)

Chapter 4
Towards Autonomically-Capable Processes: A Vision and Potentially Supportive Methods

Roy Oberhauser and Gregor Grambow

Abstract Process-aware information systems have a significant potential to support both systems and humans in various processes by partially or completely automating certain activities. However, greater adoption and inclusion are currently hindered by the cost of ownership and significant investments due to the burdens associated with the design, modeling, implementation, testing, optimization, manual adaptation, variant management, and exception handling for processes, besides the general administrative system management costs. While purely autonomic processes would ideally not require any human interaction, this chapter describes a vision for autonomically-capable processes, where processes are practically supported by the system in such a way that they reduce the human interaction burdens by being context-aware and exhibiting certain self-configuration, self-adaptation, self-optimization, and self-healing capabilities. Various cross-cutting aspects and issues requiring consideration in order to address these challenging capabilities are discussed, and potentially useful methods and techniques towards achieving the vision are highlighted. Furthermore, an example system targeted at one domain that exhibits these capabilities to some degree is used to illustrate how a combination of various techniques can be synergistically applied to a system to support incremental autonomic capabilities for processes.

4.1 Introduction

An increase in the application, utilization, and reliance on computer-controlled systems can be observed throughout society for both automation purposes as well as assisting and guiding human activities. This is illustrated, among others, by the

R. Oberhauser (✉) · G. Grambow (✉)
Computer Science Department, Aalen University, Aalen, Germany
e-mail: roy.oberhauser@hs-aalen.de

R. Oberhauser · G. Grambow
Institute for Databases and Information Systems, Ulm University, Ulm, Germany
e-mail: gregor.grambow@uni-ulm.de

© Springer International Publishing AG 2017
G. Grambow et al. (eds.), *Advances in Intelligent Process-Aware Information Systems*, Intelligent Systems Reference Library 123,
DOI 10.1007/978-3-319-52181-7_4

increased interest in the Internet of Things [1], Industry 4.0 [2], and Assisted Living [3, 4]. Undesirably, a correlation with an increased degree of unwanted interaction and administrative burden with such systems [5] can also be observed, often reflected in increasing information technology (IT) administration costs or labor.

To address this administrative burden problem, autonomic computing (AC) has articulated a self-management goal for computing systems [6, 7]. With regard to intelligent process-aware information system (PAIS), the application of such AC self-management objectives to process properties is pertinent. From the user or management perspective, a major advantage for utilizing an intelligent PAIS is not necessarily that it exhibit "smart" artificial intelligence-based process decisions. Rather, it is that it reduces the overall human interaction and management burden currently associated with such systems, supporting the fulfillment of process objectives within given constraints via the application of some combination of possibly non-intelligent techniques.

Thus, this chapter focuses on enabling autonomic properties in *processes* in contrast to only the PAIS as an IT system itself. We refer to these processes then as *autonomically-capable processes* (ACP). In supporting ACP at the *process* granularity level, ancillary or indirect AC benefits toward self-management on the PAIS IT system level may result, such as reducing IT administrator overhead or optimizing PAIS-required system resources. The scope of this chapter is not exhaustive, but rather the intent is to present a vision of ACP, discuss associated issues, and highlight selected related work as examples of methods and technologies that may be useful in incrementally achieving the ACP vision. Familiarity with concepts and methods for enabling flexibility in PAIS, as discussed in [8], is assumed.

Section 4.2 provides some background on AC concepts. An ACP vision and expected AC capabilities are presented in Sect. 4.3. Towards achieving this vision, Sect. 4.4 describes cross-cutting aspects and potential issues that require consideration, together with selected concepts, techniques, and technologies that may be applicable and potentially supportive to achieving ACP. A domain example in Sect. 4.5 shows how an adaptive PAIS can be extended with a combination of techniques for incremental support of certain ACP properties. This is followed by the chapter's conclusion.

4.2 Background on Autonomic Computing

Autonomic computing (AC) aims to equip today's complex system topography with self-management capabilities to unburden users and administrators as well as to improve security [6]. Autonomic systems can be characterized by their self-X or self-* properties, with the most well-known stemming from IBM's autonomic initiative [7] shown in Fig. 4.1.

The increasing functionality and scope of autonomic control is likely to be incremental and can be viewed as a long-term evolution in IT systems. When an observer of a system, such as a user or administrator, perceives that the capabilities exhibited

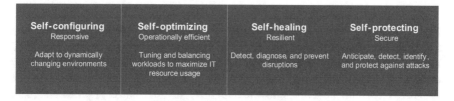

Self-configuring	Self-optimizing	Self-healing	Self-protecting
Responsive	Operationally efficient	Resilient	Secure
Adapt to dynamically changing environments	Tuning and balancing workloads to maximize IT resource usage	Detect, diagnose, and prevent disruptions	Anticipate, detect, identify, and protect against attacks

Fig. 4.1 Common autonomic computing attributes (adapted from [7])

Fig. 4.2 IBM's MAPE-K (adapted from [7])

exceed some threshold and reduce the interactions below some minimum, the system may already be regarded by them as an "intelligent system". When in fact no undesired interactions with the system occur, while fulfilling the observer's expectations, then the system may be perceived as being fully autonomous.

IBM's AC architecture involves touchpoints that implement sensor and effector behavior for some to-be-managed resource, mapping sensor and effector management interfaces to existing interfaces. Autonomic managers are components that manage other software or hardware components via touchpoints using a feedback control loop that includes monitor, analyze, plan, and execute functions utilizing knowledge (MAPE-K, see Fig. 4.2). The monitor function collects details and correlates them to symptoms, the analyze function determines if a change request is necessary, the plan function generates an appropriate change plan, while the execute function schedules and enacts the change plan. Shared knowledge is obtained or created and used by these functions.

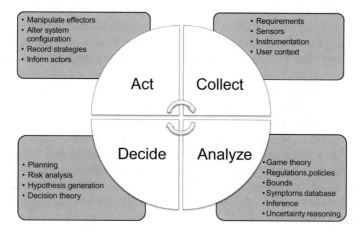

Fig. 4.3 Techniques for feedback control loops in autonomic systems (adapted from [9])

As feedback control loops play a significant role for any self-adaptive system, Fig. 4.3 shows possible types of techniques that may be utilized for equivalent functions [9]. The techniques shown are exemplary, and various techniques can be helpful and utilized for each step, while other techniques may be appropriate for certain systems. Usually the first step—collect, involves collecting data from, for instance, environment sensors, instrumentation, the application requirements, and user context. In analyze, the collected data is analyzed, which can involve checking against regulations, rules, policies, given bounds, envelopes, and a symptoms database, or using inference, game theory, uncertainty reasoning, or economic models. For decide, planning, risk analysis, hypothesis generation, and decision theory could be applied. In the act step, effectors may be manipulated, the system configuration altered, the users and administrators informed, and the strategies recorded for later analysis.

Additional self-X properties are possible. Organic computing [10], a biologically inspired form of autonomic computing, includes such properties as self-explanation and self-description [11] as well. Other properties in this field include self-awareness, goal-orientation, and self-adaptation, with some properties overlapping others. Further contemporary approaches to this topic include [12–14].

4.3   A Vision for Autonomically-Capable Processes

The ultimate vision of autonomic processes in relation to intelligent PAIS may be that the system just "knows" or anticipates what processes a user or system desires or needs and makes it happen. Yet for the foreseeable future, humans—in the role of process designers or modelers, will be involved in specifying such aspects as the process goal(s), process constraints (e.g., which resources are allowed, authorizations required, hard and soft activity dependencies), and the process priorities for which a process should be optimized (efficiency, reliability, etc.).

Since processes have a different lifecycle than systems, they pose special challenges regarding adaptation due to their wide and potentially unlimited process configurations, situations, and human interactions. When attempting to map self-X capabilities to processes rather than to systems, differences arise. For instance, the autonomic concept of self-optimization is typically understood with respect to system-level IT resources, but processes affect other real-world resources as well. Since self-protection must be considered at the PAIS system level, its support at the process level will likely reflect and depend on how the autonomic capability is supported within a particular product (e.g., available authorization and authentication mechanisms, auditing). In contrast to completely automated background processes, human-centric ACP may expect and exhibit certain user interactions and still be considered autonomic. Additionally, a process instance is not wholly responsible for performing self-X changes itself, but it is understood to perform such changes in conjunction with functionality contained in the PAIS in which it operates, with the PAIS acting on a process as an agent.

4.3.1   Vision

What we envision by autonomically-capable processes is that they exhibit *self-configuration*, *self-adaptation*, *self-optimization*, *self-healing*, and *context-awareness*, minimizing unnecessary hindrances and manual intervention to the degree considered acceptable by the process user, modeler, or designer. Briefly, given minimal input and direction by humans:

- process configurations are generated partially, as in process segments, or completely, for instance, by suggesting a process configuration to the process designer that can satisfy a given goal (*self-configuration*);
- process instances and/or process schemas are automatically and flexibly adjusted to changes in machine or (human) agent context or choices (*self-adaptation*);
- a process instance or schema is appropriately managed or evolved when exceptional events occur (*self-healing*);
- each of these areas is continually optimized to achieve improvements in line with overall objectives for the processes (*self-optimization*); and
- the PAIS system has knowledge and awareness of the current state and changes to external and internal context and takes such context into account (*context-awareness*).

Note that while capabilities typically involve the ability to perform or do something, we extend this here to also include the ability to *know* something, since it is an inherent property required for autonomic performance. Thus, we view and include *context-awareness* as a key prerequisite capability for the above self-X capabilities, due to both the special challenges and plethora of environments and unique systems in which processes operate, and due to the impacts and needs of ACP in process design and modeling.

4.3.2   Challenges

These as yet unfulfilled capabilities we also designate as *challenges for ACP*. They are viewed here in relation to the functional support provided for processes, rather than in relation to the PAIS and the IT environment it inhabits. These capabilities are mapped to a PAIS as illustrated in Fig. 4.4. Their meaning in the context of autonomic processes is further elucidated here.

Self-configuration of processes: this capability implies that, given certain goals and constraints by a process modeler at build-time, the system is able to fully configure or model a process schema itself, or provide configuration assistance to a process modeler. It does so by structuring and sequencing the process activity elements necessary to achieve the process goals while fulfilling the constraints to the greatest possible extent with regard to the given priorities. The self-configuration may involve

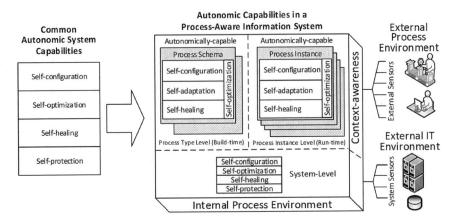

Fig. 4.4 Mapping autonomic computing properties to a PAIS

self-optimization of the process schema if it not only solves the planning, but also improves the planning toward some given priority metric.

Self-adaptation of processes: primarily this capability involves the dynamic adaptation or self-reconfiguration of a running process instance, reconfiguring it to the changing situation. It implies contextual awareness to have sufficient knowledge to adapt itself and assumes an initial configuration exists on which the adaptation occurs. If the adaptation is applicable to the type and not just to the instance, it can involve schema adaptations and thus be viewed as schema evolution. Any automated change to a process instance or process schema, e.g., a process instance reconfiguration or schema evolution, can be viewed as self-adaptation. Note that for AC, self-configuration is often understood to be self-adaptation. For ACP this is differentiated, since the initial configuration of a process model involves a different set of competencies and human interaction than does the enactment of an adaptation.

Self-healing processes: this capability corresponds to the automatic discovery of and appropriate handling of (un)anticipated process exceptions. In the case of anticipated process exceptions, the process has access to a foreseen process handler to deal with the exception. In the case of unanticipated process exceptions, an appropriate corrective handling approach must be determined. While self-healing may also involve some type of self-adaptation of the process instance, it differs in its intention, that being corrective action to deal with a process exception.

Self-optimization of processes: this capability is an automated adaptation that optimizes or improves some process metric, be it at build- or run-time. At build-time, self-configuration may involve self-optimization of the process schema if it not only solves the planning (initial configuration), but also optimizes the planning toward some given process metric. At runtime, self-optimization is an adaptation with the intent of improving some metric and may affect only the process instances or the process schema as well if it applies to all of these process types. Optimization is a form of adaptation that differs mainly in its intention or effect. Not all runtime adaptations may be optimizations, but a runtime optimization will involve some form of adaptation. Essentially, the running process instance and/or the modeled process schema may be optimized via adaptations based on certain process metrics.

Context-aware processes: this is a prerequisite capability for the above self-X capabilities. From the MAPE–K viewpoint, it involves <u>M</u>onitoring the external and PAIS-internal environment via sensors, <u>A</u>nalysis of significant contextual change, and sufficient <u>K</u>nowledge to understand both the meaning of events and the environmental, process, and system state. In the case of a PAIS, this will likely involve a domain-specific external sensor and event framework to gather the process context.

While *self-protection* can also be considered an important autonomic capability at the process level, it will necessarily be tightly integrated with PAIS system-specific protection mechanisms, and is thus out of scope for this chapter.

4.4 Achieving Autonomically-Capable Processes

Because PAIS are often integrated in diverse and complex environments, this section describes a number of potentially relevant cross-cutting aspects and/or issues to be considered from various perspectives in conjunction with each ACP capability. Furthermore, selected sources for potentially related work that may help to address an aspect or support a capability are mentioned for further reading. These aspects or issues are not intended to be sufficient, exhaustive, or definitive, but are provided as a starting point or aid to understand, discuss, and evaluate techniques or approaches potentially applicable to ACP.

Many of the hitherto typical process *perspectives* [14], used in specifying and modeling a specific process—function, behavior, information, organization, operation, and time, will likely involve more extensive consideration and complexity when involving ACP. Furthermore, additional perspectives may be necessary or helpful. Since for ACP the process context will often have unique considerations for a process, including the various environmental sensors, events, and data flows, as well involving their semantic meaning in the form of knowledge models, it too can be considered as a perspective in the specification and modeling of an ACP. Thus, we explicitly add a separate *context process perspective*.

Noting that crosscutting aspects can apply to multiple process perspectives as well as ACP capabilities, it can be useful to view such aspects in separate dimensions and consider the issues they involve at their intersection. This is analogous to how the Zachmann Framework [15] can be applied for the analysis of IT systems. Using a 3-dimensional matrix concept, in Fig. 4.5 *perspectives* are shown as rows, crosscutting *aspects* are shown as columns, and ACP *capabilities* are shown in the depth dimension. This figure serves as a conceptual reference for the discussion that follows. These aspects and perspectives and their differentiation are not intended to be complete or dogmatic, but rather show that in dealing with ACP complexity, these various areas may intersect and require separate consideration, and that such a technique can be useful.

Fig. 4.5 - Aspects and perspectives affecting self-management capabilities of autonomic processes

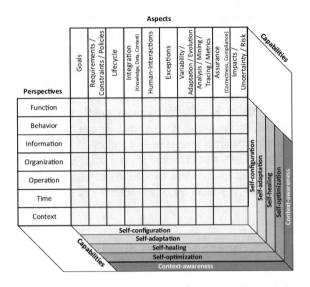

In the following discussion, the well-known process perspectives (with the addition of context) will be retained, while aspects are used to group and highlight crosscutting issues that affect both perspectives and capabilities.

4.4.1   Aspects Affecting ACP

Aspects and issues that affect ACP in a crosscutting way across process perspectives and ACP capabilities include: process goals, requirements/constraints/policies, process lifecycle, integration (knowledge, data, context), human interactions, exceptions, change (variability, adaptation, evolution), analysis/mining/tracing/metrics, assurance (correctness, compliance), and impacts/uncertainty/risk.[1] These are not intended to be comprehensive, but are rather indicative of possible considerations to be taken into account for ACP, since such considerations can affect one or more capabilities across one or more process perspectives.

4.4.1.1   ACP Goals

While traditional PAIS may have some process goals documented in a natural language for humans, with ACP these goals shall also be formulated for use by computational agents while remaining understandable and analyzable by humans.

---

[1]Many of the attributes that apply to self-adaptive systems also apply to ACP, and the so-called "modeling dimensions" [17] and associated questions, modified for ACP, will be referred to in certain aspects later.

These goals may involve business, functional, performance, safety, compliance, and other areas, and the potential evolution, flexibility, duration, multiplicity, and dependency of these goals must be considered. The notion and differentiation of hard- and soft-goals may also apply [16], implying some form of explicit prioritization.

Consider the self-adaptive dimensions from [17] for *Goals* (evolution, flexibility, duration, multiplicity, dependency): Are the ACP goals static or dynamic? Can the goals evolve, change, increase, or decrease in number during the lifetime of the process or system? How should this be specified or constrained? How should potential conflicts between multiple goals be handled? How much uncertainty or flexibility is specified in a goal (rigid, constrained, or unconstrained)? Is the goal persistent or temporary (short, medium, long) in duration? How many goals apply to what areas and how do they relate to each other (complementary, conflicting, or independent)?

For specifying ACP goals, examples of possible techniques that might be leveraged include Goal-Oriented Requirements Engineering in the form of KAOS [18, 19], or Risk-Driven Requirements Engineering [20]. As to how such goals could be integrated in a computational form for ACP, Jadex [21] provides an example of how such support can be integrated into a belief-desire-intention (BDI) model, with [22] showing how it can be combined with business process management (BPM). In the area of semantic BPM, [23] describes a business goal ontology and links these to process models. Weaknesses and issues in the currently available modeling notations are discussed in [24], and ACP may well require the development of a new and unambiguous modeling notation.

4.4.1.2 ACP Requirements, Constraints, and Policies

In contrast to ACP goals, this aspect is concerned primarily with limitations. ACP self-adaptation and self-optimization will need access to requirement and constraint specifications that will likely involve multiple process perspectives. For self-configuration, declarative, constraint-based, or similar loosely specified process modeling approaches [14] would also require the specification of requirements and constraints in order to produce valid process structures. [17] makes the point for self-adaptive systems that requirements will need to be formulated differently (e.g., with new requirement languages and methods) to explicitly deal with uncertainties and eventualities while supporting the ability to engineer assurances. Modeling notations must be created or enhanced and standardized to support the advanced specification that will be required for ACP.

Related areas from AC that may provide inspiration include the Autonomic System Specification Language [25], which provides language features for specifying autonomous parts of a system, integrating both validation and code generation facilities. The Service Component Ensemble Language [26] provides an example of a language-based approach for specifying autonomic components and interactions.

4.4.1.3 ACP Lifecycle

The traditional process lifecycle [27] consists of:

- a *design* phase where management determines the business objectives and requirements of a process;
- a *model* phase where the concrete model with its elements is configured as a schema using some notation;
- an *execute* phase where the model is enacted as a process instance and exception handling and unanticipated process changes are treated;
- and a *monitor* phase where tracing and execution analysis or diagnosis occur.

The process lifecycle in adaptive PAIS [28] was typically assumed to be manually adapted. While manual adaptation can still occur in ACP, automation will change the adaptive process lifecycle as shown in Fig. 4.6. Here dashed boxes were added to indicate where ACP capabilities initiate or assist in various PAIS lifecycle steps, dashed circles indicate computational agents who supplement human agents in influencing an ACP, and dashed lines indicate additional interactions. Certain self-adaptations or self-optimizations may affect only one process instance, multiple instances, or may apply to the schema for all new instances. Self-configuration will assist the human during design and modeling phases to remove the burdensome configuration where possible. Self-healing will decide on appropriate handlers and adjust instances accordingly when exceptions occur. Context-awareness will require environmental instrumentation and integration across the various process phases. The lifecycle will become temporally compressed, with rapid configuration or modeling changes being reflected in new or existing processes, with the monitoring and analysis of processes running concurrently rather than sequentially. Variant management will likely also play a role in the lifecycle.

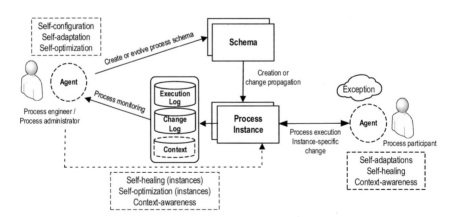

Fig. 4.6 Adaptive PAIS lifecycle augmented for autonomically-capable process support (adapted from [28])

Lifecycle issues that potentially arise in ACP scenarios are as follows: when does or should a process configuration re-enter the modeling phase? When does a redesign become necessary (inhibiting further new process instances until the redesign is resolved), or when would human intervention in the design be recommended (notify humans, but still allow for new process instances)? Should a redesign adjust the current process instances or cause them to be terminated and replaced? Under what circumstances should a new instance be triggered/enacted, an old instance be suspended or be terminated, or an adaptation be attempted? Under what circumstances should any or all running process instances be terminated? Is termination possible without side effects? How many variants should be tolerated? Under what circumstances should a process be considered a "zombie" (not making progress) and be retired? How will thrashing (over-adaptation) be detected and handled?

4.4.1.4 Integration of Knowledge, Data, and Context in ACP

As with other autonomic systems, ACP also require domain- and process-specific knowledge in order to automatically adjust processes appropriately. Knowledge has its own lifecycle in distinction to the process lifecycle, and this can create issues with regard to knowledge utilization in ACP: How does one determine what knowledge is required, sufficient, and relevant in a specific domain? How are the sources for this knowledge modeled, integrated, and accessed? How are dynamic changes to this knowledge addressed? How are the effects and dependencies between knowledge and processes addressed and verified (so that knowledge changes do not unbeknownst break an ACP)? How will humans be integrated in managing this autonomic knowledge?

While many knowledge-based management concepts and systems exist [29], the lack of standardization will make the modeling and lack of easy reuse of the knowledge more burdensome. Examples of approaches integrating knowledge and BPM systems include [30]. Case-based reasoning (CBR) has been applied as a knowledge management technique in processes [31], and [32] integrates change semantics when combining CBR with an adaptable PAIS to support process learning and evolution.

Data flows, their integration as input and output, and their quality and resiliency will play a significant part in supporting the automation of ACP. If humans are to be less burdened with ACP interaction, the ACP will need access to a larger spectrum of data and data history in order to have the necessary basis for analysis and for making plans and decisions. Various technologies (e.g., web services) have made the integration of data simpler. The self-configuration and self-adaptation in ACPs will require an enhanced awareness of the possible data flows and how they should or could be integrated in the process models. COREPRO [33] provides an example of runtime adaptation of data-driven process structures.

Context modeling and its incorporation in all process lifecycle phases (design, modeling, enactment, monitoring) will be critical for achieving ACP capabilities. Since only a subset of the context can be sensed, what states or events will be relevant for the ACP, and what can be explicitly ignored? What aspects of the context are applicable to which process perspectives, what sensors with what capabilities are available for gathering what aspects of context? Which integration mechanisms must be provided? Which sensor or event thresholds matter? Which changes to context affect which process segments? Since typical PAIS products have tended to be general and not domain-specific, while context appears to be quite domain-dependent, the integration of context presents issues. Perhaps various domain-specific context frameworks will be provided, and some type of integration or modeling standardization in this area would support pluggable context frameworks into PAIS products. [34] discusses related research issues in this area.

4.4.1.5 Human Interactions and Human-Centric ACP

With regard to humans, one can foresee a spectrum of ACP types, from highly interactive user-driven (or human-centric) processes that provide limited focused guidance to completely automated service processes where no human interaction is expected. Nevertheless, an ACP must be able to deal with any human intervention when it occurs. ACP will change how humans interact with processes and it will in turn be affected by evolving process expectations.

One part of this aspect is responding to human inputs: For instance, while users of an adaptive PAIS are responsible for making appropriate adaptations, ACP will exhibit self-adaptation. However, human interaction and influence should still be allowed. Thus, a human adaptation should not necessarily be immediately undone by an autonomic process, i.e., giving the human the impression the ACP thinks it "knows better." However, if the context has changed significantly over time for a human-based adaptation made in a quite different context, it may be appropriate to revisit it or allow for the automated adaptation to that area. Should all manual changes remain durable and "off-limits" for automated changes forever, perhaps blocking certain automated change types? Alternatively, does a significantly changing context exceeding some threshold thus permit an ACP to adapt or undo human changes? Thus the intention, competency, experience, and reliability of the human who made the change in an assessable form, and not just the process change itself with its cause, dependencies, and context state, may need to be accessible to ACP.

A further part of this aspect is soliciting human inputs: under what circumstances are human interactions required and from whom? Under what circumstances should an ACP request input from humans? When, by not soliciting assistance, will an ACP take the risk of not making progress towards its objectives, and what lies within and beyond the influence of human intervention? How should the interaction and communication with users occur, do they require insight or visualization of the process models or not? Another issue is determining when there is any benefit to notifying and bothering a human at all, since they too may not be able to enable the process to succeed given certain circumstances. In other words, when should human escalation not be invoked and just fail or give up in order to have the benefits of ACP?

References [35, 36] provides users with abstractions from technical process specifications using views. To improve comprehension, [37] abstracts parts of the model using semantic similarity of activities using structural information of the models, while [38] applies a two-phase procedure for aggregating parts of process models. While these techniques improve model visualization, they do not abstract and simplify the communication with users.

4.4.1.6   Exceptions in ACP

Since greater automation and less human intervention will be expected from ACP, more thought will be required into the types of anticipated and unanticipated exceptions that may occur across each of the various perspectives of a particular ACP and how these might be addressed or handled appropriately. This may involve specifying explicit handlers for certain exceptions (from sending notifications to specifying alternative functions), invoking strategic handlers for certain exception types, assessing the utility of past approaches, or automatically adapting the process appropriately. Some of the current issues regarding exception patterns and exception handling are described in [39, 40]. Metrics can also be used to assess the likelihood of errors in process models or improve their correctness [41], and thus may be relevant when addressing exceptions.

4.4.1.7   Change (Variability, Adaptation, and Evolution) in ACP

The change aspect considers variability, adaptation, and evolution. Variability includes the type and the points in the process (such as control flow) where change is allowed or expected, and allows for or excludes (i.e., all process instances must be alike) process variants. Adaptation may involve expected changes, perhaps specifying process fragments to be used or applied under certain conditions, and exactly what type of change is allowed. Evolution involves some series of process adaptations over time.

Variability can only be supported at certain points in a process structure that are allowed or expected to vary, known as *variation points* [42], and these must be identified for an ACP. Additionally, any constraints and dependencies on the degree and type of adaptation that can be applied to a variation point must be specified.

Exactly what types of adaptations are necessary or possible and how these are supported will be a key question as to the sufficiency and degree of flexibility support for ACP. A taxonomy of process flexibility, for instance, any shown in [14], can be useful in assessing the degree of autonomic adaptation achieved. Given an automated support for adaptation and the integration of frequently changing environmental context, one can assume that there will be an accompanying increase in the number of process variants. This may entail PAIS architectural changes to scale process variant support, as well as require human support mechanisms for navigating and analyzing the increased number of process variants.

The questions to consider for this aspect, in view of the self-adaptive dimensions from [17] for *change* (source, type, frequency, and anticipation), are as follows: Is the origin of a (potential) change external or internal? Is it functional, non-functional, or technical in nature? Is the change frequent or rare? Can the change be foreseen (taken care of), foreseeable (planned for), or unforeseen (not planned for)?

Considering the self-adaptive dimensions from [17] for *mechanisms* (type, autonomy, organization, scope, duration, timeliness, and triggering): Is the type of adaptation related to process parameters, or is it structural, or both? What degree of intervention is necessary or desired (autonomous to assisted)? What entity organizes the adaptation or is responsible it (centralized, decentralized)? Is the scope of the adaptation global or local (e.g., to the process schema or only a process instance)? What is the duration of the adaptation? For how long may the process be suspended? How timely can the adaptation occur, and will it occur soon enough to remain applicable? On which event, time trigger, and parameter threshold should an adaptation be based?

With regard to supporting self-optimization in process instances, process instance variants will need to be managed as well as all adaptations made, and any process instance differences and the associated process outcomes tracked.

Self-adaptations or self-optimizations of a more strategic nature will be referred to as *process evolution* [32], and will have a focus on the strategic change of ACP schemas. Automated process variant mining in combination with process metrics can be utilized to help improve the ACP schemas over time.

4.4.1.8   Analysis, Mining, Tracing, and Metrics for ACP

This aspect primarily deals with the past—the *historical* record and measurements of an ACP, and the ability to utilize this information both manually and automatically for improvements. As there are multiple possible use cases for utilizing historical process information, these are grouped under this aspect. Activities that make use of this data include the *monitoring* and *analyzability* [14] of the performance of an ACP—meaning that that which is needed for a certain analysis can be and has been captured, the *accountability, provenance,* and *traceability* of any adaptations or influences, manual as well as automated *process mining* to learn from the past, *debuggability* (a type of analysis) of an ACP to localize a problem or fault, and the ability to create or specify relevant *metrics* or key performance indicators that can support self-optimization.

There will be an increasing need for assurance or understanding how and why a process performed either as intended or went awry. One can imagine that regulatory, law, or legal situations will require explanations for how or why an ACP performed some action. In view of the many variables involved, significantly more data and logging will be involved, and additional analysis functionality, to not only be able to ascertain when, what, and which agent changed a process, but why some action or change plan was invoked. For ACP, a historical record of the many variable states and events involved will be necessary. The verboseness and degree of logging could conceivably depend on and be tuned by the potential impacts and risks involved. Provenance technologies [43] could play a part in supporting this.

One can conceive of a *replay* capability or *simulation* of context based on the trace that would assist in process analysis or a post-mortem investigation. COREPRO$_{Sim}$ [44] is one example of such a process simulation tool. With ACP, *process debugging* will incur additional complexity due to the likely increase in the frequency, type, and degree of adaptations and their association to context.

Automated *process mining* techniques will be likely be involved in ACP self-optimization, but require extensions for integrating context or context mining. This, in turn, may be supportive for manual human analysis and understanding of ACP changes. ProM [45] is a well-known example, but most of its analysis tools are graphical user interface centric, which creates an issue for automated integration and utilization in ACP.

Measurements of the process models via *metrics* will likely be required for self-optimization. They may be helpful in supporting and choosing any self-configuration, and can support self-healing by helping to monitor and detect issues. While a number of metrics are known [41], others are cognitive in nature and few, if any, are standardized. Moreover, research on automated calculation and evaluation will be necessary.

4.4.1.9 Assurance of ACP

While the previous subsection dealt with the historical record, this aspect is primarily future-oriented in providing some guarantee or assurance, based on evidence, that an ACP will satisfy its specified functional and non-functional properties during operation, maintain syntactical correctness across adaptations, and meet its compliance specifications.

While much work has been done towards verifying process soundness in process configurations and adaptations (e.g., [46, 47]), because ACP operates without the element of human supervision it will likely require additional process verification capabilities to assure that adaptations are correct, the process is sound, and it meets process objectives and constraints.

Process compliance techniques such as SeaFlows [48] and C3Pro [49] may become more applicable and necessary as adaptations become automated. Process compliance will need to be integrated in the various self-X ACP procedures, while minimizing the burden for specifying process compliance constraints in such a way that automated and ongoing ACP compliance checking can occur. Further, mechanisms for determining and specifying which actions to take when non-compliance is determined—an automated action or human intervention—will be needed. Automated analysis of non-compliance will be necessary, differentiating different degrees of non-compliance and their impacts or threats and potential escalation procedures. To minimize human intervention, low-level disparities with little impact might be allowed to self-heal. However, if compliance is not improved over time, then, for instance, escalation procedures might be invoked, such as human notification or stopping an ACP and awaiting human intervention.

4.4.1.10   Impacts, Uncertainty, and Risk in ACP

This aspect is concerned with explicitly articulating the potential negative and positive *impacts* (via effects or non-effects) or consequences of a particular ACP, the manner and degree of any *uncertainty* or probability related across the various perspectives, and assessing *risks* (or in its positive form opportunities). Additionally, uncertainties applicable to each specific process perspectives and ACP capability can be explicitly assessed.

Considering the self-adaptive dimensions from [17] for *effects* (criticality, predictability, overhead, and resilience), relevant issues are as follows: What is the impact if an adaptation should fail (harmless, mission-critical, or safety-critical)? What degree of predictability (non-deterministic to deterministic) is associated with the consequences of an adaptation? What overhead (negative impact) may be associated with an adaptation (e.g., insignificant, noticeable, possible thrashing, failure)? How likely is the system (or process) to remain resilient and trustworthy when changed? How much change (frequency and degree or volume) can it withstand without significant negative impacts?

[50] raises issues of risks and risk management in the context of BPM, and could conceivably be helpful for ACP as well.

4.4.2   ACP Capabilities

Having raised a number of challenges and issues that the ACP vision faces, this subsection now mentions selected methods and technologies from various areas that can be conceivably beneficial in achieving the ACP capabilities that were shown in the depth dimension in Fig. 4.5. The aspects discussed in the previous Sect. 4.4.1, the broad application of PAIS to processes in many different industries (e.g., medical, industrial, call centers), the spectrum of process coverage (human-centric, complex, completely automated), and the differences in PAIS systems lead us to the view that, apart from overly abstract concepts, no single autonomic approach or technique is currently sufficient to achieve self-X process capabilities on a concrete level.

Thus, a dependence on various techniques that can be utilized to support the evolution of *incremental autonomicity* in PAIS will be requisite. This subsection should be viewed as a synopsis to highlight several interesting options and might be used as a starting point for further investigation. The categorization of research work below should be construed as loose, since for certain works there is potential overlap across multiple areas, and others may see their own or other work from a different standpoint. Unmentioned interesting areas and work potentially relevant to the ACP vision were not intentionally ignored, but space constraints limit our discussion. Broader holistic approaches are discussed initially, and then each capability and selected possible approaches are discussed.

Autonomic system concepts: For autonomic systems, [51] provides a survey that identifies various concepts that can be used to build autonomous systems. Another survey, focusing on self-adaptation via feedback loops, is provided by [52].

[53] provides a Markov-chains adaptation approach for AC. Thus, by a quantitative analysis, various parameters of that system can be auto-adjusted.

Autonomic workflow management systems: these must be able to dynamically adapt or reconfigure workflows or processes. [54] introduces a conceptualization of different levels of autonomy for PAIS. In [55], the autonomic workflow engine can reconfigure itself to match different workload scenarios and recover from failures. In AgentWork [56], rules are used to automatically adapt workflows when exceptions occur. SmartPM [57] dynamically adapts a running process for unanticipated exceptions without requiring any predefinition of an exception handling strategy at design-time. The approach presented in [58] enables the autonomic management of process execution on a hybrid computing infrastructure using different cloud services. Therefore, it applies different managers for different tasks, such as estimating the duration or monitoring the resources. [59] applies the autonomic MAPE model (monitoring, analysis, planning, and execution) to enable adaptations to long-running scientific workflows. The approach of [60] is also situated in this domain, enabling workflow dynamicity and autonomic capabilities based on event-condition-action rules.

Figure 4.7 illustrates that a number of techniques, perhaps used in combination, can be potentially supportive towards achieving one or more autonomic capabilities, and this is elucidated in the following.

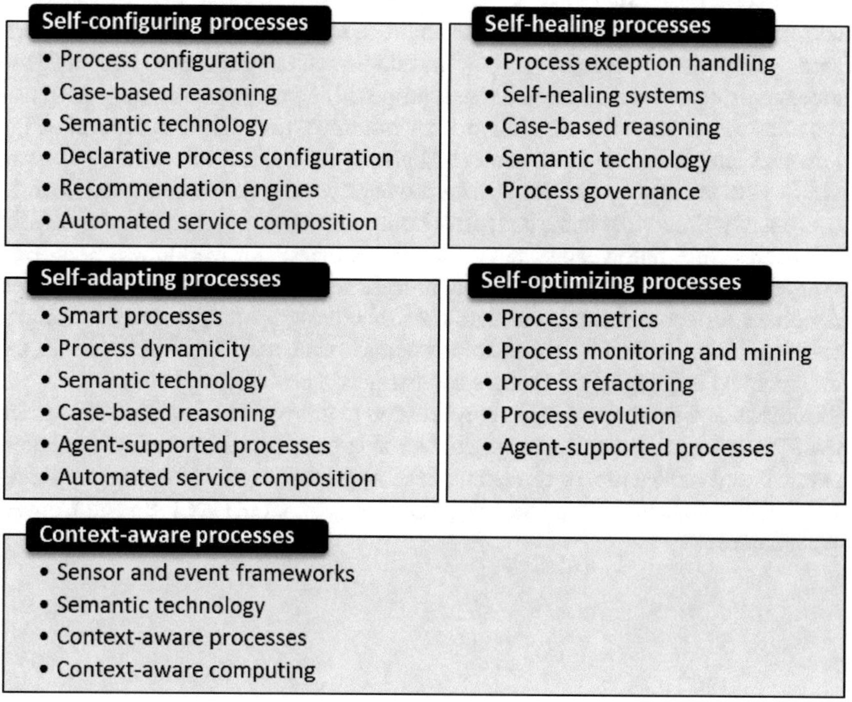

Self-configuring processes
- Process configuration
- Case-based reasoning
- Semantic technology
- Declarative process configuration
- Recommendation engines
- Automated service composition

Self-healing processes
- Process exception handling
- Self-healing systems
- Case-based reasoning
- Semantic technology
- Process governance

Self-adapting processes
- Smart processes
- Process dynamicity
- Semantic technology
- Case-based reasoning
- Agent-supported processes
- Automated service composition

Self-optimizing processes
- Process metrics
- Process monitoring and mining
- Process refactoring
- Process evolution
- Agent-supported processes

Context-aware processes
- Sensor and event frameworks
- Semantic technology
- Context-aware processes
- Context-aware computing

Fig. 4.7 Potentially supportive techniques towards ACP

4.4.2.1   Context-Aware Processes

Context-awareness has been addressed and utilized by various systems and processes, and leveraging these or similar techniques could potentially support ACPs in achieving the necessary degree of context-awareness.

```
┌─ Context-aware processes ──────────────┐
│                                        │
│  • Sensor and event frameworks         │
│  • Semantic technology                 │
│  • Context-aware processes             │
│  • Context-aware computing             │
└────────────────────────────────────────┘
```

Sensor and event frameworks: [61] details a sensor-based context framework for process-oriented information logistics. An example of a domain-specific sensor and event framework is Hackystat [62], which provides within the software engineering domain a set of sensors that can be integrated into various tools and generate events. Complex event processing [63] can also be used to generate high-level events with enriched semantic value based on low-level event pattern combinations.

Semantic technology: modeling the meaning and relationships of contextual concepts can be beneficial for achieving context-awareness, and thus semantic technologies lend themselves to supporting this capability. [64] provides one example of using semantic technology for supporting context-awareness.

Context-aware processes: Several approaches propose a way of modeling that integrates processes and context-awareness. inContext [65] primarily supports context-aware team processes. In [66], the addition of context related knowledge to processes using a so-called context tree is proposed. The context tree integrates business rules for contextually aligning process instances. [67] proposes the realization of context integration based on contextual graphs for a scientific workflow system. In [68], a system is proposed that utilizes modeled context factors to compose workflows out of web services in the domain of home assistance services in health-care. In [69], context integration and context-aware human tasks for workflows are utilized to support failure management in factory processes. In [70], a framework is proposed that allows for context-aware business process redesign. An approach for declaratively specifying context for autonomic systems is shown in [71], which extends the PerLa context language. [72] surveys context-aware process adaptation.

Context-aware computing: Surveys related to this topic include [73–75], which identify or define different factors for context-aware systems, like specific properties, context-based service use, or context models. Approaches with a focus on providing technical solutions include CASS [76], SOCAM [77], CORTEX [78], and Context Management [79].

4.4.2.2   Self-configuring Processes

Various techniques can assist in the (semi-)automated configuration and planning of workflows, reducing the amount of effort required by a human to model and structure a workflow that can achieve its ACP objective.

```
┌─ Self-configuring processes ────────────┐
│  • Process configuration                │
│  • Case-based reasoning                  │
│  • Semantic technology                   │
│  • Declarative process configuration     │
│  • Recommendation engines                │
│  • Automated service composition         │
└──────────────────────────────────────────┘
```

Process configuration: One technique is called hiding and blocking as described by [80]. For example, by blocking this approach can disable the occurrence of a single activity or event. Another way to enable process model configuration for different situations is to incorporate configurable elements into the process models as described in by [42, 81]. One example for this is a configurable activity that can be used in different ways (e.g., with a surrounding XOR pattern) or even omitted. In contrast, ADOM [82] builds on software engineering principles and allows for the specification of guidelines and constraints with the process model. Another approach is Provop [83], which enables process variants by storing a base process model and pre-configured adaptations to it. The latter can be related to context variables to enable the application of changes matching different situations. The configuration for a specific situation must be applied and is performed primarily manually. One approach is to abstract the variant configuration for the user by providing a questionnaire whose answers are mapped to certain process configuration steps [81, 84]. Another way to abstract variant management is provided by feature diagrams [85] that originated from software product line management. These diagrams offer a structured way to describe the common and the varying parts of an item. However, these approaches only offer manual configuration options that are not sufficient for the dynamically changing situations in human-centric projects.

Case handling: Case handling with regard to processes continues to be investigated [86] as a goal-oriented and flexible approach for partitioning and sequencing work in a more goal-oriented way. An example for such an approach is the one mentioned in [86] and [27]. [86] postulates a model for case handling, which is relatively data-centric and concentrates less on the control flow. Other approaches include object-aware or artifact-based process support. For more information, see the survey by [87].

Semantic technology: An example for using ontologies for process configuration is the METEOR-S project, where semantic web services [88] rely on ontologies to configure processes. [89] surveys available semantic process annotations and process composition approaches.

Declarative process configuration: DECLARE [90, 91] and Alaska [92] are two contemporary examples for declarative process management systems. Both apply declarative modeling based on constraints. This enables an arbitrary sequence of executed activities as long as no constraint is violated. DECLARE relies on YAWL [93] to execute imperative processes, and ProM [45] for log analysis and recommendations. Optional constraints are not enforced, and humans receive text messages to determine what to do. Models can be verified against dead activities, conflicting constraints, or history-based errors when altered during execution.

Recommendation engines: [94] combine a recommendation engine with a PAIS to provide user recommendations for a process, and a similar approach could conceivable be useful in supporting a human process modeler during self-configuration or in improving ACP in human-centric processes. These and similar approaches may be used for adaptation, healing, and optimization as well [95].

Automated service composition: serving as one form of process configuration, multiple services could be orchestrated or configured via automated service process techniques into a composed service or process service. AgFlow [96] is an example of a quality-of-service-aware middleware supporting quality-driven web service composition. [97] and [98] provide surveys of various web service composition methods.

4.4.2.3   Self-adapting Processes

Various techniques can assist in the automated adaptation of processes to changing situations in order for them to achieve their ACP objective.

Self-adapting processes
- Smart processes
- Process dynamicity
- Semantic technology
- Case-based reasoning
- Agent-supported processes
- Automated service composition

Smart processes: Smart processes typically include some form of self-adaptation and/or self-healing. For example, SmartPM [57] is a process management system that incorporates various techniques for reacting to unforeseen situations. That way it can recover from exceptions by applying automated adaptations. The framework in [99] enables the integration of contextual properties into workflow modeling for ubiquitous computing scenarios via the notion of the context-sensitive activity. The framework Jabbah [100] enables the integration of BPM with intelligent planning techniques to support knowledge workers, and has been tested prototypically in the e-health and the e-learning domains. The approach presented in [101] presents an integration pattern for smart workflows to support the integration of various systems in such workflows. That way, different technical processes shall be supported.

Process dynamicity: Before dynamic processes eventually support self-adaptation, they must initially support dynamic change to the imperative workflows that implement these processes. Examples technologies for evolving and dynamic processes include WASA2 [102] and ADEPT2 [103], which enable context-based process dynamicity (for instance, by a human actor) to process situations unforeseen in the process model. [104] provides another example focusing on the implementation aspects for adaptive workflow management.

Semantic technology: In SeaFlows [105], semantic constraints with regard to process compliance are taken into consideration for static validation as well as to support semantic validation of state and structural process changes.

Case-based reasoning (CBR): [106], CBR-WIMS [107] and CBRFlow [108] illustrate the application of CBR to workflows. Deeper integration and further automation of such techniques are potentially promising towards achieving the ACP vision. [27] combines adaptive process management technology with CBR to holistically support the process cycle and better handle deviations. The approach, however, lacks inclusion of context information to support automated triggering and actions.

Agent-supported processes: As an example in this area, Agentwork [56] enables automated process adaptations applied by autonomous software agents.

Automated service composition: Processes with Adaptive Web Services (PAWS) [109] is an example of a Web Services Business Process Execution Language (WS-BPEL) framework that supports self-adaptation, self-optimization, and self-healing of processes. [110] is an example of a declarative approach to self-adaptive service orchestrations.

4.4.2.4   Self-healing Processes

The autonomic system property of 'self-healing' corresponds to the appropriate handling of (un)anticipated process exceptions. Self-healing is a type of process adaptation for detected process anomalies, anticipated and unanticipated process exceptions caused by internal process states or transitions, or some explicit process constraint violation. Self-healing differs from self-adaptation primarily in its intention, that being corrective action to deal with a process exception in order for the process to still achieve its ACP objective. Various techniques can be supportive towards this end.

> **Self-healing processes**
> • Process exception handling
> • Self-healing systems
> • Case-based reasoning
> • Semantic technology
> • Process governance

Process exception handling: Process exceptions can arise for reasons such as constraint violations, deadline expiration, activity failures, or discrepancies between the real world and the modeled process [111]. The simplest case for such process exception handling is conventional workflow systems, such as the fault handlers in Business Process Execution Language engines [112] or the exception handling facility of AristaFlow [113]. However, these facilities are rather basic and only applicable for exceptions directly relating to problems in the workflows executed by these systems, and often require manual intervention.

An automated solution should be capable of process exception handling in such a way that occurring exceptions do not deteriorate process performance. Automated exception handling implies automated detection of exceptions that, in turn, depends on the capabilities of the system managing the processes [114]. However, existing PAIS are still rather limited regarding the detection and handling of exceptions [39].

A more sophisticated approach is proposed in [115]. Here, specific execution transition diagrams are used to explicitly model exceptional cases, whereby services can be withdrawn in a controlled fashion in case of exceptions. To deal with process exceptions, Agentwork [56] enables automated process adaptations applied by autonomous software agents.

An example of a smart application capable of coping with exceptional situations is SmartPM [57]. Another exception handling approach is presented in [116], where recovery strategies are proposed for process fragments that still preserve their flexibility. In turn, the approach presented in [117] enables advanced exception handling. In particular, it applies a model-based exception handling approach enabling the repair of the process and its activities. The handler even assesses repairability by analyzing the process structure and defined repair actions.

The approach presented in [118] specifically deals with exception and fault recovery for scientific workflows executed on global grids. [119] aims to create an autonomic grid workflow system that is able to adapt to changes in a grid environment. The WS-Diamond self-healing architecture [117] for service-based processes, in turn, involves the Self-healing Business Process Execution Language [120], which is based on the Processes with Adaptive Web Services. [121] integrate self-healing into the ActiveBPEL engine.

Self-healing systems: A survey of self-healing systems and their applied approaches is provided by [122], and includes approaches such as agent collaboration, replaceable agents, extensible agents, reflection, checkpointing, join points and advices, resource coordination, shared redundant information, BPEL compensation handler, supervision framework with rule violation detection, and quality-of-service and service level agreement violation detection and response.

Case-based reasoning: in addition to the case handling and CBR approaches mentioned in the previous Sects. 4.4.2.2 and 4.4.2.3, [123] provides an example of the application of CBR for self-healing in autonomic systems.

Semantic technology: one example using this technique is [124], which describes a formalism for semantic validation of semantically annotated processes that can automatically suggest bug fixes.

Process governance: the governance of processes involves some form of evaluation and control of processes, ensuring that the processes comply with certain guidelines, policies, and rules aligned with the process objectives and organizational interests. Self-healing processes may utilize process compliance checking to detect compliance discrepancies. These could then trigger an appropriate adaptation that may bring the process back into compliance, or notify responsible parties of a process issue. In relation to ACP assurance, Sect. 4.4.1.9 referred to certain process compliance approaches. These and other compliance checking techniques, like those mentioned in [14], could be used to detect a process non-compliance state—which could serve as a trigger for an autonomic self-healing action. [125] provides a comparative evaluation of regulatory compliance management in BPM, which includes compliance enactment and the use of automated compliance recovery and/or violation resolution.

4.4.2.5 Self-optimizing Processes

Optimization is a strategic process change that results in a measured improvement in an ACP's ability to achieve its objective, meaning it results in an adaption but differs mainly in the triggering (e.g., using some process metric threshold), intention (adapt to improve), and expected result of a change (e.g., improvement of some metric). Self-optimization can thus utilize similar techniques already mentioned and referenced for configuration, adaptation, and healing in Sects. 4.4.2.1, 4.4.2.2, and 4.4.2.3, including various process configuration techniques, declarative process configuration, recommendation engines, automated service composition, smart processes, process dynamicity, process-oriented case handling, CBR, agent-supported processes, process exception handling, and self-healing systems.

> **Self-optimizing processes**
> - Process metrics
> - Process monitoring and mining
> - Process refactoring
> - Process evolution
> - Agent-supported processes

Process metrics: [126] describes various process metrics and how these can be used in conjunction with a process mining tool. Additional metrics might include performance, memory usage, exceptions, errors, or anomalies.

Process monitoring and mining: tools and techniques for analyzing and mining process execution logs can be categorized as primarily for discovery, conformance, or extensions [127]. Process change logs include information on applied changes and change transactions. [14] provides further references and detail about how these process mining techniques can be used, while [128] provides a survey of semantic process monitoring and mining techniques.

Process refactoring: example approaches on how to detect process similarities and refactoring opportunities include [129, 130].

Process evolution: [14] discusses various process schema evolution and process instance migration techniques.

Agent-supported processes: [131] discusses various process optimization techniques, some of which could be leveraged by agents to automatically optimize processes.

4.5   Towards ACP: A Hybrid Extension Approach Example

The ACP vision may seem complex, abstract, and challenging, and achieving the ACP vision may appear daunting, requiring the application of a number of different approaches and techniques. This section illustrates how the application of a combination of techniques previously mentioned in Sect. 4.4 in conjunction with available off-the-shelf PAIS technology can support *incremental*[2] ACP-like *capabilities* in a concrete process-aware setting. Note that it is assumed in this section that various perspectives and crosscutting aspects that were shown in Fig. 4.5 are also considered and addressed as necessary in the actual modeling of any concrete domain-specific process, and these will no longer be explicitly mentioned.

As an illustrative domain, software engineering (SE) involves specialized knowledge workers collaborating in processes to develop or maintain a software product. To provide these software engineers with context-sensitive automated and adaptive process-oriented guidance, the Context-aware Software Engineering Environment Event-driven frameworK (CoSEEEK) was created. As shown in Fig. 4.8, CoSEEEK synergistically combines a number of different paradigms and techniques in order to provide holistic process support with ACP-like capabilities [132], summarized in a simplified manner as follows.

Semantic web computing, with its formal structuring of information and machine-processable semantics, has the potential to heterogeneously support standardized ontologies using for instance the Web Ontology Language to precisely define the semantic meaning of the domain-specific concepts. *Service-oriented computing*, with its reliance on web services, provides platform-neutral tooling integration for arbitrary applications *Space-based computing* is a powerful paradigm for coordinating autonomous processes by accessing tuples (an ordered set of typed fields) in a distributed shared memory (called a tuple space) via messaging, thereby exhibiting linear scalability properties and minimizing shared resources. *Multi-agent computing* or multi-agent systems support autonomous and interactive collaborative behaviors. *Event-based computing* allows the flow of the software functionality to be determined

Fig. 4.8 CoSEEEK's multi-paradigm approach [132]

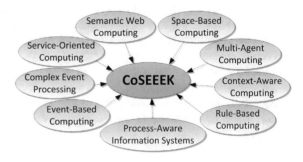

---

[2]By incremental autonomic support we mean improvements or partial support towards a certain ACP capability.

Fig. 4.9 CoSEEEK
technical architecture

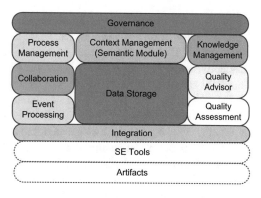

by events, supporting context-awareness with temporal data and allowing reactive
and proactive behaviors. *Complex event processing* [63] or event stream processing
is a concept to deal with meaningful event detection and processing using pattern
detection, event correlation, and other techniques to detect complex events from simpler events. *Context-aware computing* is concerned with the acquisition of context
(e.g., using sensors to perceive a situation), the abstraction and understanding of
context (e.g., matching a perceived sensory stimulus to a context), and application
behavior based on the recognized context (e.g., triggering actions based on context)
[133]. In *rule-based computing*, a collection of rules is applied to a collection of
facts via pattern matching via algorithms. *Process-aware information systems* separate process logic from application code while avoiding data- or function- centricity.
While workflow management systems [134] can be viewed as an enabling PAIS
technology, a key feature of PAIS is to support process change [109, 135, 136].

   CoSEEEK's logical architecture is shown in Fig. 4.9. Extensible and exchange-
able communication between the different components is facilitated by using events
stored in the *Data Storage* component. In addition to providing a space for decou-
pled interaction, *Data Storage* also provides extensible markup language (XML)
and relational storage. The integration of CoSEEEK with its environment is realized via an *Event Processing* component that supports the automatic acquisition and
processing of events from SE tools using web service-based sensors. Context is modeled and acquired by a *Context Management* component, which relies on semantic
technologies. To integrate the data with process execution and extend this with additional knowledge, the *Context Management* component is tightly integrated with a
Process Management component, which is in charge of process execution. The latter
component also manages dynamic adaptations to processes to conform to changing
situations. Tightly integrated with the *Context Management, Knowledge Management* centrally manages knowledge to enable comprehensive knowledge support and
provisioning for entire projects. *Quality Assessment* analyzes metrics to determine
appropriate responses, and *Quality Advisor* entails communication and interaction
with users via a graphical user interface. The *Collaboration* component detects event
patterns. A multi-agent system and a rule engine integrate configurable automatisms
into the framework to support users in their complex tasks.

To elaborate on the *Process Management* component, it realizes process enactment and automation by wrapping and abstracting the interaction with the AristaFlow BPM Suite [14]. AristaFlow, based on ADEPT technology, provides flexible support of adaptive and dynamic processes [137]. In a plug-and-play like fashion, new process templates can be composed out of existing application services, serving as schema for the robust and flexible execution of related process instances. During enactment, selected process instances can be dynamically and individually adapted in a correct and secure way; e.g., to deal with exceptional situations or evolving business needs [138]. Adaptation patterns supported by AristaFlow include the dynamic insertion, deletion, or movement of single process activities or entire process fragments respectively [139]. Integration of change functions and other services is supported via the AristaFlow Open API [140]. Dynamic process instance changes can be conducted at a high level of abstraction, hiding complexity related to dynamic process instance changes (e.g., correct process schema transformations, correct mapping of activity parameters, state adaptations). Finally, AristaFlow provides techniques for evolving process schemes [28].

This section will now briefly summarize how CoSEEEK combined and applied selected techniques introduced in Sect. 4.4 (shown in bold in Fig. 4.10) towards incrementally realizing ACP-like capabilities.

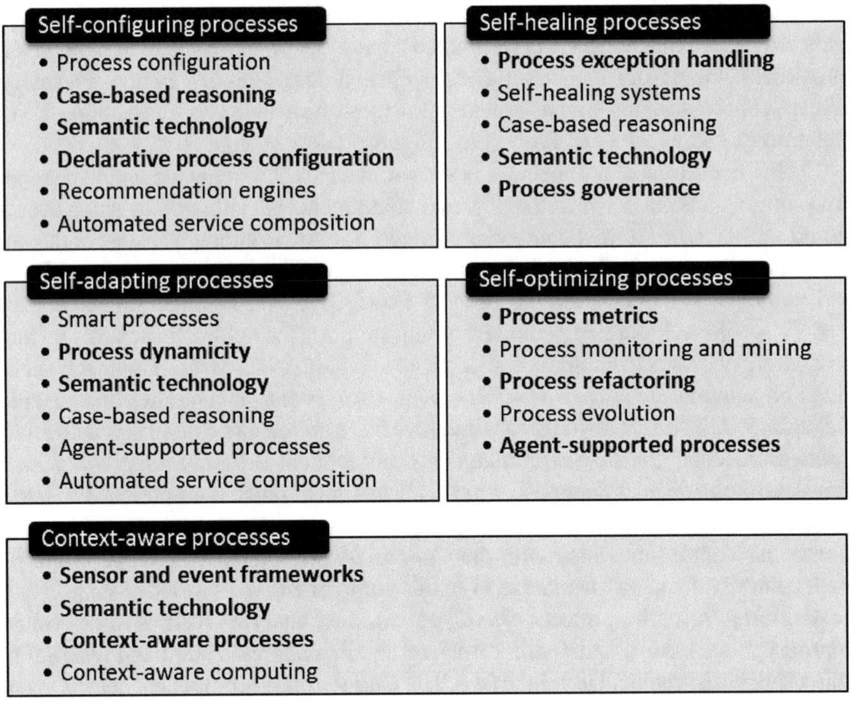

Self-configuring processes
- Process configuration
- **Case-based reasoning**
- **Semantic technology**
- **Declarative process configuration**
- Recommendation engines
- Automated service composition

Self-healing processes
- **Process exception handling**
- Self-healing systems
- Case-based reasoning
- **Semantic technology**
- **Process governance**

Self-adapting processes
- Smart processes
- **Process dynamicity**
- **Semantic technology**
- Case-based reasoning
- Agent-supported processes
- Automated service composition

Self-optimizing processes
- **Process metrics**
- Process monitoring and mining
- **Process refactoring**
- Process evolution
- **Agent-supported processes**

Context-aware processes
- **Sensor and event frameworks**
- **Semantic technology**
- **Context-aware processes**
- Context-aware computing

Fig. 4.10 Selection (shown in bold) of potentially supportive techniques towards ACP that will be illustrated with CoSEEEK

4.5.1 Towards Context-Aware Processes

Due to the heterogeneity and frequent change inherent in SE tools and environments, achieving context-aware processes in SE is a challenge. CoSEEEK provides an example of how the context can be acquired by continually extracting and processing relevant events.

```
Context-aware processes
 • Sensor and event frameworks
 • Semantic technology
 • Context-aware processes
 • Context-aware computing
```

Sensor frameworks: A domain-specific sensor framework is utilized to aggregate basic low-level events and place them in Data Storage for current and historical analysis. Various SE tool-specific sensors automatically generate events in different situations, as, e.g., source code file versioning or switching the current view in an integrated development environment.

Complex event processing: The Event Processing component utilizes complex event processing [63], generating high-level events with enriched semantic value based on detected basic event patterns.

Semantic technology: The Context Management component utilizes semantic technology to attain awareness of the SE project, processes, tools, artifacts, users, and environmental context. The advantages of semantic technology include enhanced interoperability and reuse capabilities between different applications, as well as advanced content consistency checking [141]. It enables a vocabulary for the modeled entities including taxonomies and logical statements about the entities. Ontologies further provide the capability of reasoning about the contained data and inferring new facts.

4.5.2 Towards Self-configuring Processes

As an example towards the ACP self-configuration capability, CoSEEEK shows how situational influences from the context can be taken into account to better support self-configuration of human-centric process models, as well as how a declarative approach for constructing (i.e., configuring) the processes can be applied. Synopses and excerpts from [142–144] are used to convey the primary concepts involved.

Self-configuring Processes

- Process configuration
- Case-based reasoning
- Semantic technology
- Declarative process configuration
- Recommendation engines
- Automated service composition

Using case-based reasoning and semantic technology. Situational method engineering [145] was applied in CoSEEEK to adapt generic methods to the actual contextual situation of a project. It utilizes two influence factors: process properties, which capture the impact of the current situation, and product properties that realize the impact of the product currently being processed by that process (the type of component, e.g., a graphical user interface or database component). To strike a balance between rigidly pre-specified process models and the absence of process guidance, a generic process for each use case is then dynamically extended with activities matching the current situation. A so-called case base and a method repository are utilized in the process construction. The case base contains a process skeleton of each of the use cases. These use cases are associated with a process goal (such as an assignment for the software engineer) and an attributed process and are simply referred to as cases hereafter. The process skeleton belonging to a case only contains the fundamental activities always executed for that case. The method repository contains all other activities whose execution is possible according to the case, shown as B and C in the example in Fig. 4.11. To be able to choose the appropriate activities for the current artifact and situation, the activities are connected to properties that realize product and process properties of situational method engineering.

The process goal, such as an assignment for a software engineer like refactoring code or bug fixing, is mapped to exactly one case relating to exactly one process skeleton. To realize a pre-selection of activities (e.g., 'Create Branch' or 'Code Review') that semantically match a case, the case is semantically connected to an activity via an n-to-m relation. The activities are connected to properties, which are concepts used to explicitly model contextual properties of the current situation and case

Fig. 4.11 A bug fixing process self-configured using situational method engineering (adapted from [144])

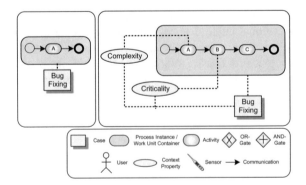

(e.g., complexity = high, urgency = low, criticality = high). The selection of an activity can depend on various process as well as product properties. To model the characteristic of a case leading to the selection of concrete activities, the case is also connected to various properties. The properties have a computed value indicating the degree in which they apply to the current situation. Utilizing the connection of an activity and a property, selection rules for activities based on the values of the properties can be specified. For further details, refer to [142].

Using declarative process configurations. Another approach towards supporting self-configuration of processes is declarative process modeling. To illustrate how this is used for process self-configuration in CoSEEEK, process activities are declaratively selected and sequenced to enable dynamic construction of the process for a process goal based on property values, provided either manually or via the context or configuration. A connection between properties and activities is utilized for this purpose, with an activity depending on one or more context or process properties. Examples include selection rules such as:

- 'Choose the activity code inspection if the risk is very high, criticality is high, and urgency is low' or
- 'Choose the activity code review if risk and criticality are both high'.

Since declarative process modeling approaches incorporate a certain amount of flexibility in the process models [146], they can adjust to different situations. However, declarative modeling can be difficult to understand and can produce models that are hard to maintain [91]. Consequently, this declarative process modeling approach uses very simple constraints and so-called building blocks that enable further structuring of the process and structural nesting.

CoSEEEK uses work unit containers to group work units corresponding with activities in a process instance. As shown in Fig. 4.12, the work unit containers are modeled above and the derived processes for execution below. 'Work Unit

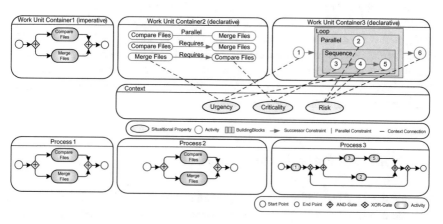

Fig. 4.12 Imperative and declarative modeling (adapted from [144])

Container 1' shows a simple, imperatively modeled process that remains unchanged for enactment (as 'Process 1').

In contrast, 'Work Unit Container 2' illustrates declarative modeling of the same process: the exact structure of the process is not rigidly pre-specified. There are only simple constraints connecting activities in the process, such as 'Requires', which expresses that one activity requires the presence of another, and 'Parallel', which expresses that both activities should be executed in parallel. The generated process for these constraints looks exactly like the imperatively modeled 'Work Unit Container 1'. Note that activities also have relations to contextual properties so that the system can contextually select a subset of the pre-specified activities for the execution process.

Furthermore, 'Work Unit Container 3' demonstrates the concept of building blocks, a grouping by ordering that supports further structuring of the process in more complex process structures. Building blocks enable hierarchical structuring of activities contained in processes and can be reused in different work unit containers readily, where they can be treated like simple activities hiding the complexity of the contained activity structure. That way, simple basic modeling is enabled while retaining the ability to model complex structures. In Fig. 4.12, three building blocks are shown for sequential, parallel, and repeated execution of the contained elements. 'Process 3' shows how a process is built based on constraints and the building blocks. It further demonstrates contextual relations, in this case assuming that the contextual properties of the situation led the system to select activities '1', '2', '3', and '5' while omitting activities '4' and '6'. An example can be activities related to software development, like coding, testing, or documenting. These can be structured by the building blocks, e.g., in a Loop enabling multiple iterations of coding, documenting, and testing combined in one building block. The latter can, e.g., be called the 'Software Development Loop' and then be easily reused as a single activity. This, in conjunction with the simple basic constraints, supports simple and understandable process models.

This approach combines the advantages of imperative and declarative modeling: The imperative processes generated for execution ensure that users follow the predefined procedures while aiding the users with process guidance. However, by declaratively specifying various candidate activities for these processes and connecting them to situational properties, the system retains the ability to choose the appropriate activities for the users' concrete situation. For more information on the concrete concepts involved, see [143, 144].

4.5.3 Towards Self-adapting Processes

Towards the ACP capability of self-adaptation of processes, CoSEEEK provides an example of using process dynamicity and semantic technology, enabling a human-centric running process instance to automatically insert appropriate activities. A synopsis and excerpts from [147] are used to convey the primary concepts involved.

```
┌─────────────────────────────────────────────┐
│ Self-adapting processes ──────────          │
│ • Smart processes                            │
│ • Process dynamicity                         │
│ • Semantic technology                        │
│ • Case-based reasoning                       │
│ • Agent-supported processes                  │
│ • Automated service composition              │
└─────────────────────────────────────────────┘
```

Leveraging process dynamicity and semantic technology. The automatic insertion of new activities into a running process instance must consider various factors to enable a syntactically correct and semantically matching insertion. That includes supplying appropriate process management and context management data for the newly inserted activity. For process executability, all variables utilized by activities for decisions in a process are supplied with initial values that can later be adjusted during execution.

The PAIS contained in the Process Management component is unaware of context but supports structural soundness checks. The Process Management component itself focuses on horizontal process governance within a process instance, and is responsible for insertion of an activity at the appropriate point in the process. For vertical governance, the Context Management component manages the connection between a work unit and work unit container. That component provides assignments, assignment activities, and atomic tasks for the complex activity that is inserted. These can be predefined when creating the processes, since the activities to be inserted also have templates from which they are created. Furthermore, the Context Management component selects a semantically matching insertion point. This is supported by the concepts of the *extension point* and the *extension*. The former is an annotation for work units that can be defined as part of their templates, and the latter is the addition. When a work unit is annotated by an extension point, the system can consider the insertion of a new activity as a direct successor of that work unit. The explicit definition of the extension points allows the system to automatically check data availability or syntactical correctness, whereas semantic suitability checks for an inserted activity requires additional information. Some SE quality actions might apply, for example, only to the end of an iteration, phase, or termination of a project. Thus, extension point feature properties can then be matched to the properties of possible extensions. That way the system can later autonomously select a suitable insertion point for a new activity. The data needed by the new activity can be predefined by template concepts.

Figure 4.13 illustrates a running process instance 'A' with activities 'A1' – 'A4' that are contextually annotated in the Context Management component in a corresponding work unit container 'A' and containing corresponding work units 'A1' – 'A4'. The complex activity to be inserted is represented by the process/work unit container 'B' with container activities/work units 'B1' – 'B4'. The activity/work unit 'B' is inserted into the running process/work unit container 'A' at extension point XP1. Further details on this approach for autonomically adapting a dynamic process instance can be found in [147].

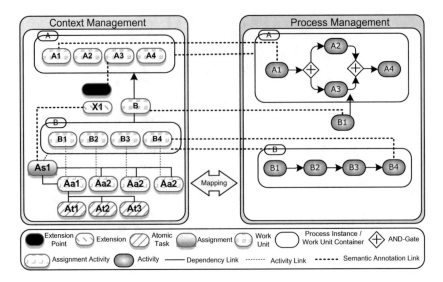

Fig. 4.13 Automated process adaptation

4.5.4 Towards Self-healing Processes

To exemplify approaches that can support self-healing, we describe a context-aware process exception handling approach and a process governance approach that can support automated corrections. Synopses and excerpts from [148–150] are used to convey the primary concepts involved.

> **Self-healing processes**
> - **Process exception handling**
> - **Self-healing systems**
> - **Case-based reasoning**
> - **Semantic technology**
> - **Process governance**

Utilizing process exception handling and semantic technology. In the highly dynamic SE domain, process exceptions can arise from various sources and be related to activities, artifacts, or the process itself, and it can be difficult to distinguish between anticipated and unanticipated exceptions. Even if they are detected, it can be difficult to directly allocate them to a simple exception handler.

Thus, our focus in extending the PAIS process exception capabilities towards supporting self-healing was to address the following requirements:

- Automatically detect the occurrence of exceptions, including the inference of an exception based on various events acquired from the environment.

- Situationally determine appropriate exception handlings based on the correct classification of the exception as well as contextual factors (e.g., properties of the current project or situation).
- Automatically determine the person responsible for exception handling in this situation, which could be the person responsible for an activity, an artifact, or perhaps the principal of a person. That does not imply that a manual intervention is necessary, but rather is used to determine which processes associated with that person may be affected and may be used for notification.
- Automatically initiate and govern exception handlings. When the parameters of the exception and the planned handling are determined, the system must have access to the process as it is enacted to automatically initiate an exception handling, distribute the handling to the responsible person or process, and govern its execution.
- Manage incomplete knowledge about exceptions. In many situations, not all data relating to an occurred exception may be available. The system shall be capable of taking action in these situations also, utilizing the incomplete knowledge.

The concept relies on contextual information, its modeling in the system, and its detection by sensors to support automated handling of the exceptions while satisfying the requirements already elicited regarding exception detection, exception handling, and the distribution to a matching human agent for handling of the exception. To apply a unified and repeatable approach to automated exception handling, we apply a set of concepts and a well-defined procedure. The latter can be roughly understood as an extended flexible variant of event-condition-action [151]. The three phases are called Recognition, Processing, and Action here. The involved abstract concepts used for exception handling are as follows:

Process exception: An exception is a deviation from the planned process and was recognized to have a potential negative impact on the process, and should thus be handled to avoid such an impact. According to [14], typically there is a distinction between anticipated exceptions, whose occurrence can be easily foreseen and unanticipated ones. Standard exception handlers can be defined for anticipated exceptions, whereas this is usually not possible for unanticipated ones.

CoSEEEK chose not to discriminate between anticipated and unanticipated exceptions, nor tie standard exception handlers to specific exceptions. Flexibility is improved through the explicit separation of events, exceptions, handling of the exceptions, responsible persons, and the point in the process where a handling is invoked. Thus, occurring events are first classified and then it is separately determined whether exceptions shall be raised, what to do with them, when to do it, and who shall do that. Additionally, the approach manages different levels of knowledge about occurring events. Depending on that level of event knowledge, it can be decided whether a more generic exception shall be raised or rather a specialized one. As stated in [111], anticipated exceptions occurring during the execution of pre-specified processes include the following categories: activity failures, deadline expiration, resource unavailability, discrepancies (between a real-world process and its computerized counterpart), and constraint violations. These can be covered by various exception types like Activity-related Exception, Artifact-related Exception,

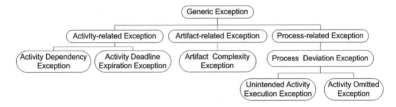

Fig. 4.14 Extract of the exception taxonomy used in CoSEEEK [148]

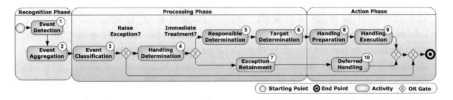

Fig. 4.15 Abstract Exception Handling Concept [148]

and Process-related Exception that can be explicitly modeled as shown in Fig. 4.14. The exception hierarchy is not intended to cover every possible exception in every project. It rather presents a basis for frequent exceptions and is extendable.

Process exception handling: The procedure for automated exception handling is separated into three phases named *Recognition*, *Processing*, and *Action* phase, each of which comprises several activities conducted by the system. The procedure will now be described with reference to Fig. 4.15 in the following, beginning with the activities of the Recognition phase.

1. Event detection: To enable automated assistance for exception handling, the detection of events related to exceptions must be automated. In a SE project, these events relate to processed activities and artifacts and thus also to supporting tools. Therefore, the Event Management component gathers a multitude of events from various tools like integrated development environments or source control management tools.

2. Event aggregation: Automatically recognized events relating to the tools in an SE project provide information about currently executed activities. Nevertheless, these events are often of rather atomic nature (like saving a file) and provide no information about the complex activity a person is performing. Therefore, these atomic events need to be processed and aggregated to derive higher-level events of more semantic value (like the application of a bug fix).

3. Event classification: Positioned within the Processing phase, event classification can be used to gain further knowledge about events or other new information to be able to find a specific handling later. The new information can also be related to the current project and its properties, like, for example, its quality goals or properties of the current situation.

4. Handling determination: used to select the appropriate countermeasure activities for a triggered exception, without deciding on the person nor the time or point in the process. Since SE exceptions are usually complex and of semantic nature, no simple rollback of the activities that caused the exception can be done. When an exception has occurred, it has to be decided when and how to take countermeasures, which also depends on the current project situation. The situation can be classified using different parameters like risk or urgency. If urgency is high, meaning there is a high schedule pressure on the project, one might decide not to address the exception immediately but to retain it for deferred handling. By using event classification, this approach can cope with different levels of knowledge about events. It may choose to retain an exception if the available knowledge about it is insufficient for immediate automatically-supported handling. Furthermore, different types of exceptions with different handlings can be connected to different events relating to different levels of knowledge. Thus, relative generalized exception handlings are also possible for situations in which only a small amount of knowledge about the exception is present.

5. Responsible determination: If it is decided to take immediate action in case of an exception, the person responsible for that action has to be determined. There can be different possibilities: For example, if an exception relating to an activity occurred, the processor of that activity can be responsible or, if an exception occurred relating to an artifact, the responsible person for that artifact (or, e.g., source code package) can also be responsible for handling the exception. There may not be a direct party responsible for each processed artifact, but responsibilities can be hierarchically structured to simplify determination of the responsible party.

6. Target determination: When the responsible party for handling the exception is determined, the structural point in the process has to be determined where the handling is applied. In certain situations, it may be appropriate to directly integrate a handling in a running process. In other cases, a separate exception handling process might be initiated.

7. Exception retainment: If, due to various parameters of the situation, no immediate handling is favored, the exception is retained in an exception list. That list can be analyzed, e.g., at the end of an iteration by the project manager.

8. Handling preparation: After all parameters for the handling of an exception are determined, the concrete handling has to be prepared, i.e., a new process instance has to be created or the handling has to be integrated seamlessly into a running process instance.

9. Handling execution: Finally, the prescribed handling is executed by the chosen agent.

10. Deferred handling: Since the exceptions are retained by the system, a list of the retained exceptions can also be presented to a human, who can decide on a handling or decide to discard certain exceptions.

In addressing ACP self-healing, this approach shows how semantic technology with process exception handling can deal with different levels of knowledge concerning events and exceptions, and thus does not require the separation between anticipated and unanticipated exceptions. The combination of environmental aware-ness with the semantic capabilities also enables the discovery of links between activ-ities and exceptions that have no direct connection. These features also support the determination of a situationally matching handling for an exception. Finally, the flexibility of the handling is enhanced by separating the determination of the han-dling, the responsible party, and the target of the handling. This approach is further described in [148].

Process governance. A number of process assessment standards applicable to SE (ISO 9001, CMMI, ISO/IEC 15504) depend on reference models and rely on the evi-dence of practices to determine if and to what degree a specific process complies with some reference model expectation. Currently this data is typically manually acquired and then correlated with expected model attributes to assess compliance. To address this, CoSEEEK utilizes an ontology-based approach to further automate process assessment while simultaneously supporting diverse process assessment reference models. It manages to unify the diverse references models with abstract concepts tied to process activities. It utilizes an assessment algorithm that supports an in-the-loop automated process assessment capability that enables process actors to receive immediate feedback on process issues as well as supporting trend and other report-ing. The Context Management module, utilizing ontologies, supports the mapping and associations between various process reference model template concepts which define expectations and thresholds (e.g., process categories, capability levels, prac-tices, scales). In the Context Management module in Fig. 4.16, concept templates shown on the left provide expected values and thresholds with the associated process schema(s). The concept instances shown on the right are also associated with a con-crete process instance, which can be used to check thresholds and compliance in the loop, invoking exception handlers when compliance is affected or triggering an appropriate notification.

Further details on this approach can be found in [149, 150].

4.5.5 Towards Self-optimizing Processes

A self-optimizing approach that automatically *optimizes* the activities in one or more processes over time based on the contextual situation was developed in CoSEEEK. Agents monitor key performance indicator (KPI) metrics, and automatically assign countermeasures for detected process issues, including the automatic refactoring of the affected process instance or schema. Synopses and excerpts from [147, 152, 153] are used to convey the primary concepts involved.

> **Self-optimizing processes**
> - **Process metrics**
> - Process monitoring and mining
> - **Process refactoring**
> - Process evolution
> - **Agent-supported processes**

Fig. 4.16 Conceptual framework for automating process assessment (adapted from [149])

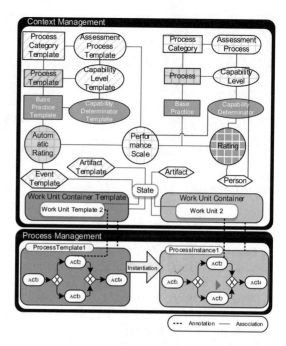

Utilizing process metrics, process refactoring, and agent-supported processes. The approach addresses the automatic detection of software quality assurance (SQA) problems and opportunities and the automatic contextual optimization and integration of SQA actions in the users' SE process. The solution addresses the following requirements:

- To be aware of problems, the system must have a facility to integrate information on process or product problems from various sources, such as analysis and tracking tools.
- To enable automated integration of quality actions at run-time, the system must be aware of quality opportunities, meaning time points when a user can cope with a quality action. This requires knowledge about the users' schedule, meaning the abstract activities that have been scheduled and estimated for the user.
- Applied quality actions should be automatically chosen during run-time in alignment with project goals in order to match the defined strategy of the project.
- Quality actions should not only rely on detected problems, but also consider common quality enhancement. Thus, proactive and reactive actions should be available.
- Context-sensitive tailoring of proposed actions is desirable considering different factors of the actual situation, e.g., properties of the applying person and application time point.
- The selection of actions should be aware of their effectiveness to optimally match specific environments or situations in different organizations. Therefore, continuous monitoring of the quality of the process output is essential to detect potential

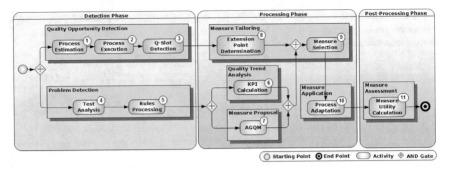

Fig. 4.17 Processing for Automated Quality Assurance (adapted from [147])

impacts of applied actions on the overall quality. In particular, a relation between the application of SQA actions and the evolution of process output should be established to assess the effectiveness of the actions.

- The automated distribution of quality actions should not interfere with standard process execution. It should be rather seamlessly integrated with other processes that are part of the process. That way, an enhanced traceability of quality action application can be fostered.
- The quality action integration should also not interfere with the users. They should not be disturbed by quality action proposals.

The approach utilizes three phases to satisfy these requirements: detection, process, and post-processing, explained in reference to Fig. 4.17.

Detection Phase. The Detection Phase enables continuous awareness of the current project situation. For integrating quality actions, two factors are of particular interest. The first factor 'Problem Detection' aggregates and analyzes report metrics (4) and rules processing determines if a threshold is exceeded (5), in which case a standard countermeasure for that problem type is assigned. The second factor 'Quality Opportunity Detection' determines the availability of opportunities for SQA actions in the users' schedule. A Q-Slot exists if the process is completing faster than expected, thus use of the slot for a SQA action would detrimentally affect the planned schedule. Process activity durations are automatically estimated (1) based on historical data. Process execution and actual activity durations are tracked (2). When a user completes an activity, the detection for quality opportunities 'Q-Slot Detection' (3) can be started.

Processing Phase. The Processing Phase deals with the selection and proposal of the quality actions. A 'Quality Trend Analysis' is performed (6) using metric KPIs, Question KPIs, and Goal KPIs. An automated form of the Goal-Question-Metric (GQM) technique [154] was developed called AGQM. The AGQM step (7) utilizes a multi-agent system with one agent assigned per project goal. Appropriate quality actions are initially proposed by each agent in alignment with its assigned project goal. Reactive actions are based on detected issues, while proactive actions "should be done" preventatively. The agents compete via bidding sessions using various strategies to determine which agent's goal and action should be prioritized. Because there are typically many known problems in a project and too many to have them all dealt with, reactive actions use a cooperative voting session to determine which action is prioritized, in essence those that help the most important quality goals. The ratio of proactive to reactive sessions can be set and vary over the life of a project, e.g., only proactive in the beginning and only reactive in the end. To prepare these actions for their automated application, 'Measure Selection' (9) incorporates information about the applying persons and the possible points in the users' schedule in which to apply the action. This leads to a selection of appropriate points (so-called Extension Points (8) and to an automated integration of the quality actions into the refactored process of the chosen person (10).

Post-Processing Phase. Finally, to be able to track the quality of the project continuously, in the Post-Processing Phase, a 'Measure Assessment' (11) is performed utilizing the quality trend analysis. This analysis supports an awareness and automatic assessment of the utility of the applied actions. Since each project is unique, the applicability and effectiveness of actions can vary with respect to different projects. Therefore, the system executes an assessment phase to rate the applied actions and to incorporate their impact in the given project and self-optimize the selection of future actions for similar situations.

Further details can be found in [147, 152, 153].

4.6 Summary

This chapter initially described the need for more autonomic support for IT-supported processes. After briefly introducing autonomic computing concepts, a vision for autonomically-capable processes was presented, describing the capabilities of self-configuration, self-adaptation, self-optimization, self-healing, and context-awareness and their connotation when appropriated to autonomically-capable process-aware information systems. In dealing with autonomic process capabilities, it described certain aspects and issues that should be considered across the common process perspectives.

Since currently no single approach, technique, or technology is likely to be sufficient, incremental degrees of increasing autonomicity can be achieved by leveraging a combination of techniques. Towards this end, for each autonomic capability, this chapter presented a selection of potentially supportive or promising methods from various research areas. Finally, an example from the software engineering domain served to practically illustrate how selected approaches and techniques can be combined within a single system to achieve a greater degree of autonomicity for the various process capabilities.

References

1. Atzori, L., Iera, A., Morabito, G.: The internet of things: a survey. Comput. Netw. **54**(15), 2787–2805 (2010)
2. Herrman, M., Pentek, T., Otto, B.: Design principles for Industrie 4.0 scenarios: a literature review. Working Paper 01/2015, Technishe Universität Dortmund (2015)
3. Rashidi, P., Mihailidis, A.: A survey on ambient-assisted living tools for older adults. IEEE J. Biomed. Health Inf. **17**(3), 579–590 (2013)
4. Deegan, P., Grupen, R., Hanson, A., Horrell, E., Ou, S., Riseman, E., Sen, S., Thibodeau, B., Williams, A., Xie, D.: Mobile manipulators for assisted living in residential settings. Autonom. Robots **24**(2), 179–192 (2008)
5. Ganek, A.G., Corbi, T.A.: The dawning of the autonomic computing era. IBM Syst. J. **42**(1), 5–18 (2003)
6. Kephart, J.O., Chess, D.M.: The vision of autonomic computing. Computer **36**(1), 41–50 (2003)
7. IBM: An architectural blueprint for autonomic computing. IBM White Paper, 3rd ed. (2005)
8. Reichert, M., Weber, B.: Enabling Flexibility in Process-Aware Information Systems: Challenges, Methods, Technologies. Springer (2012)
9. Dobson, S., Denazis, S., Fernández, A., Gaïti, D., Gelenbe, E., Massacci, F., Nixon, P., Saffre, F., Schmidt, N., Zambonelli, F.: A survey of autonomic communications. Trans. Auton. Adapt. Syst. (TAAS) ACM **1**(2), 223–259 (2006)
10. Schmeck, H., Müller-Schloer, C., Çakar, E., Mnif, M., Richter, U.: Adaptivity and self-organization in organic computing systems. Trans. Auton. Adapt. Syst. (TAAS) ACM **5**(3), 10:1–10:32 (2010)
11. Würtz, R.P. (ed.): Organic Computing (Understanding Complex Systems). Springer (2008)
12. Richter, U., Mnif, M., Branke, J., Müller-Schloer, C., Schmeck, H.: Towards a generic observer/controller architecture for organic computing. GI Jahrestagung **93**(1), 112–119. Gesellschaft für Informatik (2006)
13. Nafz, F., Ortmeier, F., Seebach, H., Steghofer, J.-P., Reif, W.: A generic software framework for role-based organic computing systems. In: Proceedings of the ICSE Workshop on Software Engineering for Adaptive and Self-Managing Systems (SEAMS'09), pp. 96–105. IEEE (2009)
14. Santambrogio, M.D., Hoffmann, H., Eastep, J., Agarwal, A.: Enabling technologies for self-aware adaptive systems. In: Proceedings of the NASA/ESA Conference on Adaptive Hardware and Systems (AHS), pp. 149–156. IEEE (2010)
15. Sowa, J.F., Zachman, J.A.: Extending and formalizing the framework for information systems architecture. IBM Syst. J. **31**(3), 590–616 (1992)
16. Soffer, P., Wand, Y.: On the notion of soft-goals in business process modeling. Bus. Process Manag. J. **11**(6), 663–679 (2005)
17. Cheng, B.H., De Lemos, R., Giese, H., Inverardi, P., Magee, J., Andersson, J., Becker, B., Bencomo, N., Brun, Y., Cukic, B., Serugendo, G.M., Dustdar, S., Finkelstein, A., Gacek, C., Geihs, K., Grassi, V., Karsai, G., Kienle, H.M., Kramer, J., Litoiu, M., Malek, S., Mirandola, F., Müller, H.A., Park, S., Shaw, M., Tichy, M., Tivoli, M., Weyns, D., Whittle, J.: Software engineering for self-adaptive systems: A research roadmap. In: Software Engineering for Self-adaptive Systems. Lecture Notes in Computer Science, vol. 5525, pp. 1–26. Springer (2009)
18. van Lamsweerde, A.: Requirements Engineering: From System Goals to UML Models to Software Specifications. Wiley (2009)
19. Espada, P., Goulão, M., Araújo, J.: A framework to evaluate complexity and completeness of KAOS goal models. In: Advanced Information Systems Engineering, pp. 562–577. Springer (2013)
20. van Lamsweerde, A.: Risk-driven engineering of requirements for dependable systems. In: Engineering Dependable Software Systems, vol. 34, pp. 207–234. IOS Press (2013)

21. Braubach, L., Pokahr, A., Lamersdorf, W.: Jadex: A BDI-agent system combining middleware and reasoning. In: Software Agent-based Applications, Platforms and Development Kits, pp. 143–168. Birkhäuser Basel (2005)
22. Burmeister, B., Arnold, M., Copaciu, F., Rimassa, G.: BDI-agents for agile goal-oriented business processes. In: Proceedings of the 7th International Joint Conference on Autonomous Agents and Multiagent Systems: Industrial Track, pp. 37–44. International Foundation for Autonomous Agents and Multiagent Systems (2008)
23. Markov, I., Kowalkiewicz, M.: Linking business goals to process models in semantic business process modeling. In: 12th International IEEE Enterprise Distributed Object Computing Conference (EDOC'08), pp. 332–338. IEEE (2008)
24. List, B., Korherr, B.: An evaluation of conceptual business process modelling languages. In: Proceedings of the 2006 ACM Symposium on Applied Computing, pp. 1532–1539. ACM (2006)
25. Vassev, E., Hinchey, M.: ASSL: a software engineering approach to autonomic computing. Computer **42**(6), 90–93 (2009)
26. De Nicola, R., Ferrari, G., Loreti, M., Pugliese, R.: A language-based approach to autonomic computing. In: Formal Methods for Components and Objects, pp. 25–48. Springer, Berlin (2013)
27. Weber, B., Sadiq, S., Reichert, M.: Beyond rigidity—dynamic process lifecycle support: a survey on dynamic changes in process-aware information systems. In: Computer Science - Research and Development, vol. 23(2), pp. 47–65 (2009)
28. Weber, B., Reichert, M., Wild, W., Rinderle-Ma, S.: Providing integrated life cycle support in process-aware information systems. Int. J. Cooperat. Inf. Syst. (IJCIS) **18**(1), 115–165 (2009)
29. Maier, R.: Knowledge management systems: Information and Communication Technologies for Knowledge Management. Springer (2007)
30. Jung, J., Choi, I., Song, M.: An integration architecture for knowledge management systems and business process management systems. Comput. Ind. **58**(1), 21–34 (2007)
31. Mansar, S.L., Marir, F., Reijers, H.A.: Case-based reasoning as a technique for knowledge management in business process redesign. Electron. J. Knowl. Manage. **1**(2), 113–124 (2003)
32. Rinderle, S., Weber, B., Reichert, M., Wild, W.: Integrating process learning and process evolution—a semantics based approach. In Business Process Management, pp. 252–267. Springer (2005)
33. Müller, D., Reichert, M., Herbst, J.: A new paradigm for the enactment and dynamic adaptation of data-driven process structures. In: Advanced Information Systems Engineering, pp. 48–63. Springer, Berlin (2008)
34. Rosemann, M., Recker, J.C.: Context-aware process design: Exploring the extrinsic drivers for process flexibility. In: The 18th International Conference on Advanced Information Systems Engineering. Proceedings of Workshops and Doctoral Consortium, pp. 149–158. Namur University Press (2006)
35. Bobrik, R., Reichert, M., Bauer, T.: View-based process visualization. In: Business Process Management, pp. 88–95. Springer, Berlin (2007)
36. Reichert, M., Kolb, J., Bobrik, R., Bauer, T.: Enabling personalized visualization of large business processes through parameterizable views. In: 27th ACM Symposium On Applied Computing, 9th Enterprise Engineering Track, pp. 1653–1660. ACM Press (2012)
37. Smirnov, S., Reijers, H., Weske, M.: A semantic approach for business process model abstraction. In: Advanced Information Systems Engineering, vol. 6741, pp. 497–511 (2011)
38. Eshuis, R., Grefen, P.: Constructing customized process views. Data Knowl. Eng. **64**(2), 419–438 (2008)
39. Russell, N., van der Aalst, W., ter Hofstede, A.: Workflow exception patterns. In: Advanced Information Systems Engineering, pp. 288–302. Springer (2006)
40. Adams, M., ter Hofstede, A.H., van der Aalst, W.M., Edmond, D.: Dynamic, extensible and context-aware exception handling for workflows. In: On the Move to Meaningful Internet Systems: CoopIS, DOA, ODBASE, GADA, and IS, pp. 95–112. Springer (2007)

41. Mendling, J.: Metrics for Process Models: Empirical Foundations of Verification, Error Prediction, and Guidelines for Correctness, Vol. 6. Springer (2008)
42. Rosemann, M., van der Aalst, W.M.: A configurable reference modelling language. Inf. Syst. **32**(1), 1–23 (2007)
43. Curbera, F., Doganata, Y., Martens, A., Mukhi, N.K., Slominski, A.: Business provenance—a technology to increase traceability of end-to-end operations. In: On the Move to Meaningful Internet Systems (OTM 2008), pp. 100–119. Springer (2008)
44. Müller, D., Reichert, M., Herbst, J., Köntges, D., Neubert, A.: Corepro sim: A tool for modeling, simulating and adapting data-driven process structures. In: Business Process Management, pp. 394–397. Springer (2008)
45. van der Aalst, W.M., van Dongen, B.F., Günther, C.W., Rozinat, A., Verbeek, E., Weijters, T.: ProM: The Process Mining Toolkit. In: Proceedings of the Business Process Management Demonstration Track (BPMDemos) (2009)
46. van der Aalst, W.M., van Hee, K.M., ter Hofstede, A.H., Sidorova, N., Verbeek, H.M.W., Voorhoeve, M., Wynn, M.T.: Soundness of workflow nets: classification, decidability, and analysis. Formal Aspects Comput. **23**(3), 333–363 (2011)
47. Hallerbach, A., Bauer, T., Reichert, M.: Guaranteeing soundness of configurable process variants in Provop. In: IEEE Conference on Commerce and Enterprise Computing (CEC'09), pp. 98–105. IEEE (2009)
48. Ly, L.T.: SeaFlows—A compliance checking framework for supporting the process lifecycle. Ph.D. thesis, University of Ulm (2013)
49. Knuplesch, D., Reichert, M., Kumar, A.: Visually Monitoring Multiple Perspectives of Business Process Compliance. In: 13th International Conference on Business Process Management (BPM 2015). Lecture Notes in Computer Science, pp. 263–279. Springer (2015)
50. zur Muehlen, M., Ho, D.T.Y.: Risk management in the BPM lifecycle. In: Business Process Management Workshops, pp. 454–466. Springer (2006)
51. Huebscher, M.C., McCann, J.A.: A survey of autonomic computing–degrees, models, and applications. ACM Comput. Surv. (CSUR) **40**(3), 1–28 (2008)
52. Brun, Y., Serugendo, G.D.M., Gacek, C., Giese, H., Kienle, H., Litoiu, M., Müller, H., Pezzè, M., Shaw, M.: Engineering self-adaptive systems through feedback loops. In: Cheng, B.H., de Lemos, R., Giese, H., Inverardi, P., Magee, J. (eds.) Software Engineering for Self-Adaptive Systems. Lecture Notes in Computer Science Hot Topics, vol. 5525, pp. 48–70. Springer (2009)
53. Calinescu, R., Kwiatkowska, M.: Using quantitative analysis to implement autonomic IT systems. In: Proceedings of the 31st International Conference on Software Engineering, pp. 100–110. IEEE Computer Society (2009)
54. Strohmeier, M., Ye, E.: Towards autonomic workflow management systems. In: Proceedings of the Conference of the Center for Advanced Studies on Collaborative research, pp. 34–37. IBM Corp. (2006)
55. Heinis, T., Pautasso, C., Alonso, G.: Design and Evaluation of an Autonomic Workflow Engine. In: Proceedings of the Second International Conference on Autonomic Computing, (ICAC 2005), pp. 27–38. IEEE (2005)
56. Müller, R., Greiner, U., Rahm, E.: AgentWork: A workflow system supporting rule-based workflow adaptation. Data Knowl. Eng. **51**(2), 223–256 (2004)
57. Marrella, A., Mecella, M., Sardina, S.: SmartPM: An adaptive process management system through situation calculus, Indigolog, and classical planning. In: 14th International Conference on Principles of Knowledge Representation and Reasoning (KR). AAAI Press (2014)
58. Kim, H., El-Khamra, Y., Rodero, I., Jha, S., Parashar, M.: Autonomic management of application workflows on hybrid computing infrastructure. Sci. Program. **19**(2), 75–89 (2011)
59. Lee, K., Sakellariou, R., Paton, N.W., Fernandes, A.A.A.: Workflow adaptation as an autonomic computing problem. In: Proceedings of the 2nd Workshop On Workflows in Support of Large-Scale Science, pp. 29-34. ACM (2007)
60. Zhang, G., Jiang, C., Sha, J., Sun, P.: Autonomic workflow management in the grid. In: Proceedings of the 7th International Conference on Computational Science (ICCS 2007), Part III, pp. 220–227. Springer (2007)

61. Michelberger, B., Mutschler, B., Reichert, M.: A context framework for process-oriented information logistics. In: Business Information Systems, pp. 260–271. Springer, Berlin (2012)

62. Johnson, P.M., Kou, H., Paulding, M., Zhang, Q., Kagawa, A., Yamashita, T.: Improving software development management through software project telemetry. IEEE Softw. **22**(4), 76–85 (2005)

63. Luckham, D.C.: The Power of Events: An Introduction to Complex Event Processing in Distributed Enterprise Systems. Addison-Wesley Longman Publishing (2001)

64. Wang, X.H., Zhang, D.Q., Gu, T., Pung, H.K.: Ontology based context modeling and reasoning using OWL. In: Proceedings of the Second IEEE Annual Conference on Pervasive Computing and Communications Workshops, pp. 18–22. IEEE (2004)

65. Dorn, C., Dustdar, S.: Sharing hierarchical context for mobile web services. Distrib. Parallel Databases **21**(1), 85–111 (2007)

66. Saidani, O., Nurcan, S.: Towards context aware business process modelling. In: Proceedings of the 8th Workshop on Business Process Modeling, Development, and Support BPMDS'07 in Association with CAISE'07), Vol. 7. Springer (2007)

67. Fan, X., Brézillon, P., Zhang, R., Li, L.: Making context explicit towards decision support for a flexible scientific workflow system. In: Proceedings of the Fourth Workshop on Human Centered Processes HCP, pp. 3–9 (2011)

68. Ardissono, L., Di Leva, A., Petrone, G., Segnan, M., Sonnessa, M.: Adaptive medical workflow management for a context-dependent home healthcare assistance service. Electron. Notes Theoret. Comput. Sci. **146**(1), 59–68 (2006)

69. Wieland, M., Leymann, F., Schäfer, M., Lucke, D., Constantinescu, C., Westkämper, E.: Using context-aware workflows for failure management in a smart factory. In: Proceedings of the Fourth International Conference on Mobile Ubiquitous Computing Systems (UBICOMM 2010), Services and Technologies, pp. 379–384. IARIA (2010)

70. Bessai, K., Claudepierre, B., Saidani, O., Nurcan, S.: Context-aware business process evaluation and redesign. In: Proceedings of BPMDS'08, pp. 1–10 (2008)

71. Schreiber, F.A., Tanca, L., Camplani, R., Viganó, D.: Towards autonomic pervasive systems: the PerLa context language. In: Electronic Proceedings of the 6th International Workshop on Networking Meets Databases (Co-located with SIGMOD 2011), pp. 1–7 (2011)

72. Smanchat, S., Ling, S., Indrawan, M.: A survey on context-aware workflow adaptations. In: Proceedings of the 6th International Conference on Advances in Mobile Computing and Multimedia, pp. 414–417. ACM (2008)

73. Bolchini, C., Curino, C.A., Quintarelli, E., Schreiber, F.A., Tanca, L.: A data-oriented survey of context models. ACM SIGMOD Rec. **36**(4), 19–26 (2007)

74. Baldauf, M., Dustdar, S., Rosenberg, F.: A survey on context-aware systems. Int. J. Ad Hoc Ubiquitous Comput. **2**(4), 263–277 (2007)

75. Kapitsaki, G.M., Prezerakos, G.N., Tselikas, N.D., Venieris, I.S.: Context-aware service engineering: a survey. J. Syst. Softw. **82**(8), 1285–1297 (2009)

76. Fahy, P., Clarke, S.: CASS—a middleware for mobile context-aware applications. In: Proceedings Workshop on Context-Awareness (held in connection with MobiSys'04) (2004)

77. Gu, T., Pung, H.K., Zhang, D.Q.: A middleware for building context-aware mobile services. In: Proceedings of the IEEE Vehicular Technology Conference (VTC), vol. 5, pp. 2656–2660. IEEE (2004)

78. Biegel, G., Cahill, V.: A framework for developing mobile, context-aware applications. In: Proceedings of the 2nd IEEE Conference on Pervasive Computing and Communication, pp. 361–365. IEEE (2004)

79. Korpipää, P., Jani, M., Kela, J., Malm, E.J.: Managing context information in mobile devices. IEEE Pervasive Comput. **2**(3), 42–51 (2003)

80. Gottschalk, F.: Configurable process models. Ph.D. Thesis. Eindhoven University of Technology (2009)

81. La Rosa, M.: Managing variability in process-aware information systems. Doctoral dissertation, Queensland University of Technology, Brisbane, Australia (2009)

82. Reinhartz-Berger, I., Soffer, P., Sturm, A.: Extending the adaptability of reference models. IEEE Trans. Syst. Man Cybern. Part A **40**(5), 1045–1056 (2010)

83. Hallerbach, A., Bauer, T., Reichert, M.: Capturing variability in business process models: the Provop approach. J. Softw. Maintenan. Evol.: Res. Pract. **22**(6–7), 519–546 (2010)

84. La Rosa, M., van der Aalst, W.M.P., Dumas, M., ter Hofstede, A.H.M.: Questionnaire-based variability modeling for system configuration. Softw. System Model. **8**(2), 251–274 (2009)

85. Schobbens, P.Y., Heymans, P., Trigaux, J.C.: Feature diagrams: A survey and a formal semantics. In: 14th IEEE International Conference Requirements Engineering, pp. 139-148. IEEE (2006)

86. van der Aalst, W.M.P., Weske, M., Grunbauer, D.: Case handling: a new paradigm for business process support. Data Knowl. Eng. **53**(2), 129–162 (2005)

87. Künzle, V., Reichert, M.: Striving for object-aware process support: how existing approaches fit together. In: 1st International Symposium on Data-driven Process Discovery and Analysis (SIMPDA'11). Lecture Notes in Business Information Processing, vol. 116, pp. 169–188. Springer (2012)

88. Cardoso, J., Sheth, A.: Semantic e-workflow composition. J. Intell. Inf. Syst. **21**(3), 191–225 (2003)

89. Lautenbacher, F., Bauer, B.: A survey on workflow annotation & composition approaches. In: Proceedings of the Workshop on Semantic Business Process and Product Lifecycle Management (SemBPM), Innsbruck, Austria, pp. 12–23 (2007)

90. Pesic, M., Schonenberg, H., van der Aalst, W.M.P.: Declare: full support for loosely-structured processes. In: 11th IEEE International Enterprise Distributed Object Computing Conference (EDOC 2007), pp. 287–298. IEEE (2007)

91. Haisjackl, C., Barba, I., Zugal, S., Soffer, P., Hadar, I., Reichert, M., Pinggera, J., Weber, B.: Understanding Declare Models: Strategies, Pitfalls, Empirical Results. Software & Systems Modeling. Springer, pp. 1–28 (2014)

92. Weber, B., Pinggera, J., Zugal, S., Wild, W.: Alaska Simulator Toolset for Conducting Controlled Experiments on Process Flexibility. In: Soffer, P., Proper, E. (eds.) CAiSE Forum 2010. Lecture Notes in Business Information Processing, vol. 72, pp. 205–221. Springer (2011)

93. van der Aalst, W.M., ter Hofstede, A.H.: YAWL: yet another workflow language. Inf. Syst. **30**(4), 245–275 (2005)

94. Barba, I., Weber, B., del Valle, C., Jiménez-Ramírez, A.: User recommendations for the optimized execution of business processes. Data Knowl. Eng. **86**, 61–84 (2013)

95. Barba, I., Weber, B., del Valle, C.: Supporting the optimized execution of business processes through recommendations. In: Business Process Management Workshops, vol. 1, pp. 135–140. Springer, Berlin (2012)

96. Zeng, L., Benatallah, B., Ngu, A.H., Dumas, M., Kalagnanam, J., Chang, H.: QoS-aware middleware for web services composition. IEEE Trans. Softw. Eng. **30**(5), 311–327 (2004)

97. Rao, J., Su, X.: A survey of automated web service composition methods. In: Semantic Web Services and Web Process Composition, pp. 43–54. Springer, Berlin (2005)

98. Strunk, A.: QoS-aware service composition: A survey. In: IEEE 8th European Conference on Web Services (ECOWS 2010), pp. 67–74. IEEE (2010)

99. Abbasi, A.Z., Shaikh, Z.A.: A conceptual framework for smart workflow management. In: Proceedings of the International Conference on Information Management and Engineering (ICIME'09), pp. 574–578. IEEE (2009)

100. González-Ferrer, A., Fdez-Olivares, J., Sánchez-Garzón, I., Castillo, L.: Smart process management: automated generation of adaptive cases based on intelligent planning technologies. In: M.L. Rosa (ed.), Proceedings of 8th Business Process Management Conference (Demo Track), vol. 615, CEUR Workshop Proceedings, pp. 28–33 (2010)

101. Wieland, M., Nicklas, D., Leymann, F.: Managing technical processes using smart workflows. Towards a Service-Based Internet: Proceedings of the First European Conference. Service-Wave 2008, pp. 287–298. Springer, Berlin (2008)

102. Weske, M.: Flexible modeling and execution of workflow activities. In: Proceedings of the 31st Hawaii International Conference on System Sciences, vol. 7, pp. 713-722. IEEE (1998)

103. Reichert, M., Rinderle, S., Kreher, U., Dadam, P.: Adaptive process management with ADEPT2 (Tool Demo). In: Proceedings of the 21st Int'l Conf. on Data Engineering, Tokyo, pp. 1113–1114 (2005)
104. Eberle, H., Leymann, F., Unger, T.: Implementation Architectures for Adaptive Workflow Management. In Proceedings of the the Second International Conference on Adaptive and Self-Adaptive Systems and Applications (ADAPTIVE 2010), pp. 98–103. IARIA (2010)
105. Ly, L.T., Rinderle-Ma, S., Göser, K., Dadam, P.: On enabling integrated process compliance with semantic constraints in process management systems. Inf. Syst. Frontiers **14**(2), 195–219 (2012)
106. Kapetanakis, S., Petridis, M., Knight, B., Ma, J., Bacon, L.: A case based reasoning approach for the monitoring of business workflows. Case-Based Reasoning. Research and Development, pp. 390–405. Springer, Berlin (2010)
107. Kapetanakis, S., Petridis, M.: Evaluating a Case-based Reasoning Architecture for the intelligent monitoring of business workflows. In: Successful Case-Based Reasoning Applications-2, pp. 43–54. Springer, Berlin (2014)
108. Weber, B., Wild, W., Breu, R.: CBRFlow: Enabling adaptive workflow management through conversational case-based reasoning. In: Advances in Case-Based Reasoning, pp. 434–448. Springer, Berlin (2004)
109. Ardagna, D., Comuzzi, M., Mussi, E., Pernici, B., Plebani, P.: Paws: a framework for executing adaptive web-service processes. IEEE Softw. **24**(6), 39–46 (2007)
110. Cugola, G., Ghezzi, C., Pinto, L.S.: DSOL: a declarative approach to self-adaptive service orchestrations. Computing **94**(7), 579–617 (2012)
111. Russell, N., van der Aalst, W., ter Hofstede, A.: Workflow exception patterns. In: Proceedings of the International Conference on Advanced Information Systems Engineering (CAiSE'06). LNCS, vol. 4001, pp. 288–302. Springer, Berlin (2006)
112. Kloppmann, M., Konig, D., Leymann, F., Pfau, G., Roller, D.: Business process choreography in WebSphere: Combining the power of BPEL and J2EE. IBM Syst. J. **43**(2), 270–296 (2004)
113. Lanz, A., Reichert, M., Dadam, P.: Making business process implementations flexible and robust: error handling in the AristaFlow BPM Suite. In: Proceedings of the CAiSE'10 Forum. Lecture Notes in Business Information Processing, vol. 72, pp. 174–189. Springer (2010)
114. Luo, Z., Sheth, A., Kochut, K., Miller, J.: Exception handling in workflow systems. Appl. Intell. **13**(2), 125–147 (2000)
115. Shang, Z.: Exception handling in smart process-based applications. In: Proceedings of the International Conference on Computer Application and System Modeling (ICCASM 2010), vol. 7, pp. 522–526. IEEE (2010)
116. Eberle, H., Leymann, F., Unger, T.: Transactional process fragments-recovery strategies for flexible workflows with process fragments. In: Proceedings of the Services Computing Conference (APSCC), pp. 250–257. IEEE Asia-Pacific (2010)
117. Friedrich, G., Fugini, M., Mussi, E., Pernici, B., Tagni, G.: Exception handling for repair in service-based processes. IEEE Trans. Softw. Eng. **36**(2), 198–215 (2010)
118. Sindrilaru, E., Costan, A., Cristea, V.: Fault tolerance and recovery in grid workflow management systems. In: Proc. International Conference on Complex, Intelligent and Software Intensive Systems (CISIS 2010), pp. 475–480. IEEE (2010)
119. Rahman, M.: An Autonomic Workflow Management System for Global Grids. In: Proceedings of the 8th IEEE International Symposium on Cluster Computing and the Grid, pp. 578–583. IEEE (2008)
120. Modafferi, S., Mussi, E., Pernici, B.: SH-BPEL: a self-healing plug-in for Ws-BPEL engines. In: MW4SOC '06: Proceedings of the 1st Workshop on Middleware for Service Oriented Computing (MW4SOC 2006), pp. 48–53. ACM, New York (2006)
121. Subramanian, S., Thiran, P., Narendra, N.C., Mostefaoui, G.K., Maamar, Z.: On the enhancement of BPEL engines for self-healing composite web services. In: Proceedings of the 2008 International Symposium on Applications and the Internet (SAINT '08), pp. 33–39. IEEE Computer Society, Washington, DC (2008)

122. Psaier, H., Dustdar, S.: A survey on self-healing systems: approaches and systems. Computing **91**(1), 43–73 (2011)
123. McSherry, D., Bustard, D.: Autonomic self-healing and recovery informed by environment knowledge. Artif. Intell. Rev. **26**(1–2), 89–101 (2006)
124. Weber, I., Hoffmann, J., Mendling, J.: Semantic business process validation. In: Proceedings of the 3rd International Workshop on Semantic Business Process Management (SBPM'08), Vol. 472, pp. 22–36. CEUR-WS Proceedings (2008)
125. El Kharbili, M.: Business process regulatory compliance management solution frameworks: a comparative evaluation. In: Proceedings of the Eighth Asia-Pacific Conference on Conceptual Modelling, Vol. 130, pp. 23–32. Australian Computer Society (2012)
126. Vanderfeesten, I., Cardoso, J., Mendling, J., Reijers, H.A., van der Aalst, W.M.: Quality metrics for business process models. BPM and Workflow Handbook, vol. 144, pp. 179–190 (2007)
127. van der Aalst, W.M.: Process Mining: Discovery, Conformance and Enhancement Of Business Processes. Springer Science & Business Media (2011)
128. de Medeiros, A.A., Pedrinaci, C., van der Aalst, W.M., Domingue, J., Song, M., Rozinat, A., Norton, B., Cabral, L.: An outlook on semantic business process mining and monitoring. In: On the Move to Meaningful Internet Systems 2007 (OTM 2007): Workshops, pp. 1244–1255. Springer, Berlin (2007)
129. Dijkman, R., Gfeller, B., Küster, J., Völzer, H.: Identifying refactoring opportunities in process model repositories. Inf. Softw. Technol. **53**(9), 937–948 (2011)
130. Weber, B., Reichert, M., Mendling, J., Reijers, H.A.: Refactoring large process model repositories. Comput. Ind. **62**(5), 467–486 (2011)
131. Vergidis, K., Tiwari, A., Majeed, B.: Business process analysis and optimization: beyond reengineering. IEEE Trans. Syst. Man Cybern. Part C: Appl. Rev. **38**(1), 69–82 (2008)
132. Oberhauser, R.: Leveraging Semantic Web Computing for Context-Aware Software Engineering Environments, pp. 157–179. Semantic Web, Gang Wu (ed.). InTech (2010)
133. Schmidt, A.: Ubiquitous Computing—Computing in Context, Ph.D. dissertation, Lancaster University, U.K. (2003)
134. van der Aalst, W.M., van Hee, K.: Workflow Management: Models, Methods, and Systems. MIT Press (2004)
135. Reichert, M., Dadam, P.: A framework for dynamic changes in workflow management systems. In: Proc. 8th International Workshop on Database and Expert Systems Applications, pp. 42–48. IEEE (1997)
136. Pesic, M., Schonenberg, M., Sidorova, N., van der Aalst, W.: Constraint-based workflow models: change made easy. In: Proceedings of the 15th International Conference on Cooperative Information Systems (CoopIS'07). Lecture Notes in Computer Science, vol. 4803, pp. 77–94. Springer (2007)
137. Dadam, P., Reichert, M.: The ADEPT project: a decade of research and development for robust and flexible process support. Comput. Sci.-Res. Dev. **23**(2), 81–97 (2009)
138. Reichert, M., Rinderle-Ma, S., Dadam, P.: Flexibility in process-aware information systems. Transactions on Petri Nets and Other Models of Concurrency vol. II. Lecture Notes in Computer Science, vol. 5460, pp. 115–135. Springer (2009)
139. Weber, B., Reichert, M., Rinderle, S.: Change patterns and change support features—enhancing flexibility in process-aware information systems. Data Knowl. Eng. **66**(3), 438–466 (2008)
140. Reichert, M., Dadam, P., Rinderle-Ma, S., Jurisch, M., Kreher, U., Göser, K.: Architectural principles and components of adaptive process management technology. In: PRIMIUM - Process Innovation for Enterprise Software. Lecture Notes in Informatics, P-151, pp. 81–97. Gesellschaft für Informatik (2009)
141. Gasevic, D., Djuric, D., Devedzic, V.: Model Driven Architecture and Ontology Development. Springer (2006)
142. Grambow, G., Oberhauser, R., Reichert, M.: Semantic workflow adaption in support of workflow diversity. In: Proc. 4th International Conf. on Advances in Semantic Processing (SEMAPRO'10), pp. 158–165. IARIA (2010)

143. Grambow, G., Oberhauser, R., Reichert, M.: Semantically-driven workflow generation using declarative modeling for processes in software engineering. In: 15th IEEE International Enterprise Distributed Object Computing Conference Workshops (EDOCW 2011), pp. 164–173. IEEE (2011)

144. Grambow, G., Oberhauser, R., Reichert, M.: Contextual generation of declarative workflows and their application to software engineering processes. Int. J. Adv. Intell. Syst. **4**(3 & 4), 158–179 (2012)

145. Ralyté, J., Brinkkemper, S., Henderson-Sellers, B.: Situational Method Engineering: Fundamentals and Experiences. Springer (2007)

146. Pichler, P., Weber, B., Zugal, S., Pinggera, J., Mendling, J., Reijers, H.A. Imperative versus declarative process modeling languages: an empirical investigation. In: Business Process Management Workshops, pp. 383–394. Springer (2012)

147. Grambow, G., Oberhauser, R., Reichert, M.: Contextual injection of quality measures into software engineering processes. Int. J. Adv. Softw. **4**(1 & 2), 76–99 (2011)

148. Grambow, G., Oberhauser, R., Reichert, M.: Event-driven exception handling for software engineering processes. In: Proceedings of the 5th International Workshop on Event-Driven Business Process Management. Lecture Notes in Business Information Processing, vol. 99, pp. 414–426. Springer (2011)

149. Grambow, G., Oberhauser, R., Reichert, M.: Towards Automated Process Assessment in Software Engineering. In: Proceedings of the 7th International Conference on Software Engineering Advances (ICSEA 2012), pp. 289–295. IARIA (2012)

150. Grambow, G., Oberhauser, R., Reichert, M.: Automated software engineering process assessment: supporting diverse models using an ontology. Int. J. Adv. Softw. **6**(1 & 2), 213–224 (2013)

151. Dittrich, K.R., Gatziu, S., Geppert, A.: The active database management system manifesto: a rulebase of ADBMS features. In: Rules in Database Systems, pp. 1–17. Springer (1995)

152. Grambow, G., Oberhauser, R.: Towards automated context-aware selection of software quality measures. In: Proceedings of the 5th International Conference on Software Engineering Advances, pp. 347–352. IARIA (2010)

153. Grambow, G., Oberhauser, R., Reichert, M.: Employing semantically driven adaptation for amalgamating software quality assurance with process management. In: Proceedings 2nd International Conference on Adaptive and Self-Adaptive Systems and Applications, pp. 58–67. IARIA (2010)

154. Basili, V.R., Caldiera, V.R.B.G., Rombach, H.D.: The goal question metric approach. Encyclopedia of Software Engineering, vol. 2, pp. 528–532 (1994)

Chapter 5
Process-Oriented Information Logistics: Requirements, Techniques, Application

Bernd Michelberger, Markus Hipp and Bela Mutschler

Abstract Enterprises are confronted with a continuously increasing data load. Examples of such data include sensor data, office files, e-mails, guidelines, and business data. In turn, this data overload makes it difficult for knowledge workers to identify information needed to perform their tasks in the best possible way. To remedy this information underload and to optimally utilize enterprise data, the latter must be aligned with business processes. In fact, process-related information and business processes are usually managed separately. On one hand, enterprise content management systems, shared drives, and Intranet portals are used fss management technology is used to design and enact business processes. With *process-oriented information logistics* (POIL), this chapter presents an approach for bridging this gap. In particular, POIL enables the process-oriented and context-aware delivery of process-related information to knowledge workers. We also present use cases and proof-of-concept prototypes to demonstrate the application and benefits of POIL.

Keywords Process-oriented information logistics · Business process management · Information management

5.1 Introduction

Market globalization has led to massive cost pressure and increased competition for enterprises. Products and services must be developed in ever-shorter cycles, and innovative ways of collaboration within and across enterprises are emerging. As

B. Michelberger (✉) · B. Mutschler
University of Applied Sciences Ravensburg-Weingarten, Württemberg, Germany
e-mail: bernd.michelberger@hs-weingarten.de

B. Mutschler
e-mail: bela.mutschler@hs-weingarten.de

M. Hipp
Institute of Databases and Information Systems, University of Ulm, Ulm, Germany
e-mail: markus.hipp@uni-ulm.de

© Springer International Publishing AG 2017
G. Grambow et al. (eds.), *Advances in Intelligent Process-Aware Information Systems*, Intelligent Systems Reference Library 123,
DOI 10.1007/978-3-319-52181-7_5

examples consider the treatment of patients in healthcare networks [1] and cross-organizational processes in automotive engineering [2].

5.1.1 Problem Statement

A major challenge for enterprises is the increasing amount of data they are confronted with [3]. Typical data include, for example, office files, e-mails, web data, process descriptions, process models, forms, checklists, best practices, and guidelines. In turn, all this enterprise data is provided using shared drives, databases, enterprise applications, Intranet portals, and process-aware information systems (PAIS). In particular, this heterogeneity of both the data and the data sources turns data management into a time-consuming, complex task [4, 5].

Employees not only need access to data, but require information; i.e. standardized and processed data provided for a specific purpose and in a specific work context [6]. In particular, selecting required information is even more time-consuming and difficult to handle than just managing data [4]. Often encountered problems in this context are incomplete or outdated information [7]. Another problem is the identification of required information to accomplish business processes in the best possible way. To cope with these challenges, an approach is needed that allows aligning process-related information (denoted as *process information* in the following) with business processes and their tasks; i.e. an approach enabling both *information-* and *process-awareness* [8, 9].

Information- and process-awareness, however, are not yet sufficient since the alignment of process information with business processes is strongly influenced by the work context of knowledge workers (i.e. process participants) [10]. For example, consider a process description: In a specific work context only selected parts of this description might be relevant for a knowledge worker. Furthermore, less experienced knowledge workers might need a more detailed process description than experienced ones. Hence, in order to effectively meet information needs, the work context of a knowledge worker must be taken into account as well, i.e. *context-awareness* must be additionally enabled (cf. Fig. 5.1) [11].

This chapter picks up this demand and suggests *process-oriented information logistics* (POIL) as an approach providing integrated information-, context-, and process-awareness (cf. Fig. 5.1). Specifically, POIL allows for a process-oriented and

Fig. 5.1 Problem dimensions

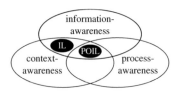

context-aware delivery of process information to knowledge workers. POIL focuses on knowledge-intensive business processes that involve large amounts of process information, expertise, user interaction, and decision-making [12].

5.1.2 Information Logistics

Traditional *information logistics* (IL) approaches deal with the question of how to deliver information to knowledge workers as effectively and efficiently as possible [13]. For this purpose, basic principles from the fields of material logistics and lean management are applied. Examples include just-in-time delivery [14] and satisfaction of customer needs [15]. Particularly, IL aims to deliver that information to knowledge workers fitting their demands best. Thus, information-awareness (e.g. awareness of information quality and flows) and, to a smaller extent, context-awareness (e.g. awareness of the user context when delivering personalized information) adopt a key role in IL [16] (cf. Fig. 5.1).

Although IL is independent from any *information and communication technology* (ICT), the latter has been intensively used as an IL enabler for several years. As examples consider ICT solutions in areas like business intelligence, management information systems, and enterprise content management. However, these solutions suffer from shortcomings including limited applicability (e.g. only applicable within an enterprise, but not across enterprises) [17], missing operational support (e.g. only the management level is addressed) [18], and lack of process-awareness (e.g. delivering information without considering the current process context). In fact, missing process-awareness in contemporary IL solutions has guided the development of POIL [19].

5.1.3 Requirements

The following requirements reflect wishes and needs of process participants who are concerned with the alignment of enterprise data and business processes (i.e. with POIL) (cf. Fig. 5.2). The requirements also concern technical issues which are necessary to enable the delivery of relevant process information to process participants.

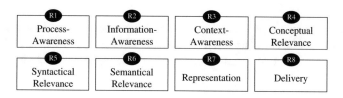

Fig. 5.2 Requirements on POIL

All requirements have been derived based on results of empirical studies and were approved by a literature survey (see [16, 20, 21] for details).

R1 (Process-awareness): POIL should be able to gather business processes from process repositories (both on the process schema and instance level), transform business processes into uniform structured process objects (on the business process element level), and integrate them into a comprehensive approach.

R2 (Information-awareness): POIL should be able to gather process information from large, distributed, and heterogeneous data sources, transform different process information into uniform structured information objects (on different quality levels), and integrate them into a comprehensive approach.

R3 (Context-awareness): POIL should be able to gather context information from sensors, transform context information into uniform structured context objects (on different granularity and quality levels), and integrate them into a comprehensive approach.

R4 (Conceptual Relevance): POIL should be able to analyze information, process, and context objects on a conceptual level (i.e. based on the topic of the object) in order to identify conceptual relationships between objects.

R5 (Syntactical Relevance): POIL should be able to analyze information, process, and context objects on a syntactic level (i.e. based on metadata of the object) in order to identify syntactic relationships between objects.

R6 (Semantical Relevance): POIL should be able to analyze information, process, and context objects on a semantic level (i.e. based on the content of the object) in order to identify semantic relationships between objects.

R7 (Representation): POIL should be able to represent information, process, and context objects as well as their relationships in a meaningful, machine-, and user-interpretable form and, moreover, in a structured way.

R8 (Delivery): POIL should be able to use information, process, and context objects to guide a process-oriented and context-aware delivery of relevant process information (i.e. information objects) to process participants.

The remainder of this chapter is organized as follows. Section 5.2 describes the main technical concepts underlying the notion of POIL. Section 5.3 deals with the application of POIL in practice. Section 5.4 discusses POIL. Finally, Sect. 5.5 concludes the chapter with a summary.

5.2 Process-Oriented Information Logistics

POIL comprises two layers; i.e. the *integration layer* and the *analysis layer*. Overall goal of these layers is to create a *semantic information network* (SIN) which represents the conceptual baseline for specific POIL applications (cf. Sect. 5.3).

5.2.1 Step 1: Integration

The integration layer integrates data from different data sources and realizes a uniform view on this data. Thereby, we distinguish between data sources of *process objects* (i.e. business processes and their tasks), *information objects* (i.e. process information), and *context objects* (i.e. context information).

Process objects correspond to process elements such as tasks, gateways or events. We thereby consider business processes both at the process schema and the process instance level. A process schema is a reusable business process template (e.g. describing patient examination processes in general) comprising, for example, tasks and sequence flows. In turn, a process instance (e.g. an examination of a certain patient) corresponds to a concrete business case that is concurrently executed with other instances of the same or other process schemas [22]. One key idea of POIL is to split up business process models into their elements (i.e. constituent process objects) as well as to integrate them with information objects (cf. Requirements R1 and R2).

In turn, *information objects* refer to process information needed when working on business processes. Examples include e-mails, office files, forms, checklists, guidelines, informal process descriptions, or best practices.

Finally, *context objects* represent context information characterizing the work context of a process participant, such as user id, roles, experiences, current tasks, used devices, locations, environment, and time [10].

Technically, for each data source, at least one *interface* has to be implemented. Interfaces transform proprietary data into generic process, information and context objects (cf. Requirements R1–R3). All generic objects follow the same structure and comprise attributes such as the id, url, author, file format, or raw content (e.g. the entire text of an e-mail or the coordinates of a user's location). The uniform object structure is a prerequisite to accomplish the conceptual, syntactical and semantical analyses required (cf. Requirements R4–R6).

The specific results of the integration are three independent object spaces: the *process object space*, the *information object space*, and the *context object space* (cf. Fig. 5.3). Each object space can be defined as a set of generic process, information, and context objects.

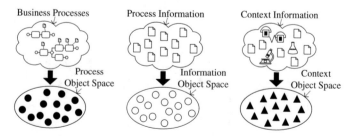

Fig. 5.3 Creation of the object spaces

Fig. 5.4 Simplified part of a
SIN

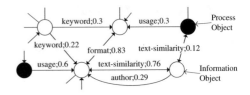

5.2.2 Step 2: Analysis

The three object spaces constitute the foundation (i.e. input) for the analysis layer.
The main purpose of this layer is to create a SIN based on the information and process
object spaces (cf. Requirement R7).

Figure 5.4 shows a simplified part of a SIN which comprises information objects
(i.e. white circles), process objects (i.e. black circles) and relations between these
objects (i.e. black arrows). Relations may exist between information objects (e.g.
a file similar to another one), process objects (e.g. an event triggering a task) or
between information and process objects (e.g. a file required for the execution of
a task). Furthermore, relations are labeled to indicate their semantics and they are
weighted to indicate their relevance. This allows determining why objects are related
and how strong their relation is (cf. Fig. 5.4).

For identifying the relations between objects, we use a combination of syntactical
and semantical analyses. These analyses are provided by and realized with a semantic
middleware [23]. More precisely, algorithms from the fields of data mining, text mining (e.g. text preprocessing, linguistic preprocessing, vector space model, clustering,
classification, information extraction), pattern-matching, and machine learning (e.g.
supervised learning, unsupervised learning, reinforcement learning, transduction)
are applied in this context [24, 25]. Algorithms are (inverse) term frequency algorithms, link popularity algorithms, and utilization context algorithms (see [19, 25,
26] for details). In summary, Fig. 5.5 shows the schematic and simplified creation of
a SIN.

In addition to the SIN, a *context model* (CM) is constructed based on available
context objects [10]. It corresponds to an ontology-based model that uses pre-defined
context factors such as user or location (see [10] for details). The CM allows capturing

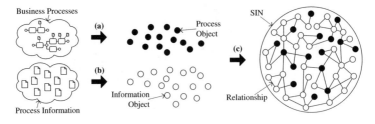

Fig. 5.5 Schematic and simplified creation of a SIN

the work context of a process participant, which can then be used to filter the SIN. Hence, the identification and delivery of the currently needed process information becomes more accurate and user-oriented; e.g. the delivery of process information can be adapted to the used device or to the experience level of a user. Note that the CM is completely independent from the SIN; i.e. context objects are solely stored in the CM, but not in the SIN. Accordingly, there exists one central SIN for all users, but a specific CM for each user.

Technically, the CM is applied to the SIN by the *SIN facade*. The latter constitutes an interface to retrieve both information objects and process objects from the SIN taking the user's current work context into account (cf. Requirement R8). We thereby distinguish between an *explicit* and *implicit* information demand. Examples of an explicit information demand include *full-text retrieval* (e.g. delivery of medical reports of a patient using the search query "John Doe report"), *concept-based retrieval* (e.g. delivery of files dealing with a certain concept like the disease "diabetes") and *graph-based retrieval* (e.g. delivery of related process information to a certain process schema) [27]. An example of an implicit information demand is *context-based retrieval* (e.g. a patient record is delivered based on the doctor's location). Therefore, the process participant's work context is applied to retrieve information and process objects.

5.2.3 The Semantic Information Network

The core component of POIL is the SIN. In general, the relationships of a SIN may exist between information objects (e.g. a guideline similar to another one), process objects (e.g. an event triggering a subprocess) or information and process objects (e.g. an instruction required for executing a task) (cf. Fig. 5.6a–c).

Further, a relationship may be either *explicit* (i.e. hard-wired) or *implicit* (i.e. not hard-wired). Explicit relationships include, for example, modeled data flows in a process schema. Implicit relationships, in turn, can be automatically identified by a variety of algorithms. They link, for example, objects addressing the same topic or objects used in the same work context [19]. Moreover, relationships are labeled and weighted. A weight is expressed in terms of a number between 0 and 1 (with 1 indicating the strongest possible relationship) [28]. This allows determining why objects are interlinked and how strong their relation is.

A SIN is a labeled and weighted *directed graph*. Each directed *edge* $e = (u, v)$ represents a relationship and is associated with an ordered pair of *vertices* (u, v),

Fig. 5.6 Relationships between objects

Fig. 5.7 Slings, parallelism
and anti-parallelism of SINs

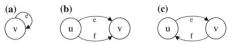

which represents information and process objects; u corresponds to the source and
v to the destination of e. A SIN can be formally defined as follows:

Definition 1 (*SIN*) A labeled and weighted digraph is called a semantic information
network $SIN = (V, E, L, W, f_l, f_w)$, iff:

- V is a set of vertices representing information and process objects
- E is a set of edges representing relationships between objects
- L is a set of labels indicating relationship reasons
- W is a set of weights representing the relevance of relationships
- f_l is a labeling function with $f_l : E \to L$
  assigning to each edge $e \in E$ a label $f_l(e) \in L$
- f_w is a weighting function with $f_w : E \to W$
  assigning to each edge $e \in E$ a weight $f_w(e) \in W$

A SIN is a *finite graph*, i.e. V and E are finite sets [29]. A SIN may contain
slings (i.e. $\exists e = (v, v)$, cf. Fig. 5.7a), *parallelism* (i.e. $\exists e = (u, v) \wedge f = (u, v)$, cf.
Fig. 5.7b), and *anti-parallelism* (i.e. $\exists e = (u, v) \wedge f = (v, u)$, cf. Fig. 5.7c).

In general, each vertex v may have several incoming and outgoing edges. The
number of incoming edges of a vertex constitutes its *incoming degree*, whereas the
number of outgoing edges is denoted as *outgoing degree*. The *total degree* of a
vertex corresponds to the sum of its incoming and outgoing degrees. Vertices having
no incoming edges are denoted as *unreferenced*. In turn, vertices without outgoing
edges are called *non-referencing*. Finally, vertices being unreferenced as well as
non-referencing are *isolated* [29, 30].

Definition 2 (*Degree*) Let $SIN = (V, E, L, W, f_l, f_w)$ be a semantic information
network. Then the number of incoming and outgoing edges of a vertex $v \in V$ is
denoted as the degree of v, where:

- $deg^-(v)$ is the incoming degree of a vertex $v \in V$ with
  $deg^-(v) = |E^-(v)| = |\{e = (x, y) \in E \mid y = v\}|$
- $deg^+(v)$ is the outgoing degree of a vertex $v \in V$ with
  $deg^+(v) = |E^+(v)| = |\{e = (x, y) \in E \mid x = v\}|$
- $deg(v)$ is the total degree of a vertex $v \in V$ with
  $deg(v) = deg^-(v) + deg^+(v) = |E(v)| = |E^-(v)| + |E^+(v)|$

Vertices directly relating to a neighbored vertex are called *internal neighborhood*,
whereas vertices referenced by another vertex are called *external neighborhood*.
The *total neighborhood* then corresponds to the union of both internal and external
neighborhood.

Definition 3 (*Neighborhood*) Let $SIN = (V, E, L, W, f_l, f_w)$ be a semantic information network. Then referencing and referenced vertices of a vertex $v \in V$ are denoted as the neighborhood of v, where:

- $\Gamma^-(v)$ is the internal neighborhood of a vertex $v \in V$ which is denoted as $\Gamma^-(v) = V^-(v) = \{u \in V^-(v)\}$
- $\Gamma^+(v)$ is the external neighborhood of a vertex $v \in V$ which is denoted as $\Gamma^+(v) = V^+(v) = \{u \in V^+(v)\}$
- $\Gamma(v)$ is the total neighborhood of a vertex $v \in V$ which is denoted as $\Gamma(v) = \Gamma^-(v) \cup \Gamma^+(v) = V(v) = \{u \in V(v)\}$

As set out in Definition 1, function f_w assigns a weight to each edge e. This weight indicates the relevance of an edge and therefore the strength of the relationship between two vertices. In a SIN, however, there may be multiple edges between vertices with different weights. In order to determine the overall strength between two vertices, we calculate the average weight of all edges between them. The *average weight* $avg_\emptyset(F)$ of a set of edges F can be calculated as follows:

$$avg_\emptyset(F) = \sum_{f \in F} \frac{f_w(f)}{|F|} \tag{5.1}$$

In practice, however, certain edges may have to be weighted higher. As an example consider a "is similar to" relationship, which is usually more important than a "has same file extension as" relationship. Therefore, in addition, we introduce significance function f_s with $f_s : E \to \mathbb{N}_1$ assigning to each edge $e \in E$ a significance value $f_s(e) \in \mathbb{N}_1$. The higher the significance value of an edge, the more important this edge will be. The *average weight* $avg_\Delta(F)$ of a set of edges F can be calculated as follows:

$$avg_\Delta(F) = \sum_{f \in F} \frac{f_s(f) * f_w(f)}{\sum_{g \in F} f_s(g)} \tag{5.2}$$

As aforementioned, we apply various algorithms provided by a semantic middleware which we use to implement the SIN in six consecutive phases (see [19] for details). These algorithms, however, do not allow identifying relevant, i.e. currently needed, information objects within a SIN. What is thus additionally needed are further algorithms. This is indispensable in order to reach the aforementioned goals of POIL, i.e. to provide users with the right process information.

5.2.4 Determining the Relevance of Process Information

In two case studies as well as an online survey [20, 21], we showed that knowledge workers spend considerable effort to handle process information. One challenging task in this context is to identify relevant process information. In POIL, the SIN constitutes the basis for this task. However, additional techniques are needed to

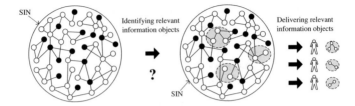

Fig. 5.8 Delivering relevant information objects

determine relevant process information, i.e. currently needed information objects in a SIN dependent on the work context (cf. Fig. 5.8).

In the following, we introduce two algorithms for identifying relevant information objects in a SIN. The first one determines the *link popularity* of information objects based on the SIN's relationship structure. The second one determines the *rate popularity* of information objects based on user ratings. Note that the algorithms can be used independently, but can be also combined.

Determining Link Popularity Usually, in enterprises, process information is not explicitly linked to other process information or business processes. Therefore, it is not possible to take advantage of a rich relationship structure within an enterprise environment. Instead, process information is implicitly linked to other process information and business process models, e.g. dealing with the same topic or used in the same process context. In particular, a SIN makes such implicit relationships explicit through its edges. The SIN's relationship structure enables the use of algorithms to identify strongly linked and therefore popular objects. As a particular challenge, however, existing link popularity algorithms are not sufficient in this context. Thus, we extend them and introduce the *SIN LP algorithm*, which allows determining the link popularity of information objects in a SIN (cf. Fig. 5.9).

Basic to any link popularity algorithm is an *InDegree algorithm* [31] that measures the *link popularity* $LP(v)$ of an information object v by taking its number of incoming edges into account (cf. Formula (5.3)). The higher the number is, the greater the popularity of an information object will become:

$$LP(v) = deg^-(v) \qquad (5.3)$$

In a SIN, the InDegree is not really helpful since certain relationships might be more valuable than others. In turn, this issue is picked up by the *PageRank algorithm* [32]: Relationships originating from information objects of high quality are considered being more valuable than relationships originating from information objects of

Fig. 5.9 Link popularity algorithms

low quality (cf. Formula (5.4)). Thus, the link popularity $LP(v)$ of an information object v can be calculated as follows (with d corresponding to a damping factor ranging from 0 to 1):

$$LP(v) = (1 - d) + d \sum_{w \in \Gamma^-(v)} \frac{LP(w)}{deg^+(w)} \qquad (5.4)$$

However, like the InDegree, the conventional PageRank (originally designed for the web) is not applicable to a SIN, since it only considers single relationships. In a SIN, there are multiple, weighted and labeled relationships. Hence, we must extend PageRank. First, we have to support multiple relationships:

$$LP(v) = (1 - d) + d \sum_{w \in \Gamma^-(v)} |\{e = (w, v) \in E\}| * \frac{LP(w)}{deg^+(w)} \qquad (5.5)$$

To support weighted relationships, we extend Formula (5.5) by including an average weighting function $avg_{\emptyset}$ (cf. Sect. 5.2.3):

$$LP(v) = (1 - d) + d \sum_{w \in \Gamma^-(v)} avg_{\emptyset}(\{e = (w, v) \in E\}) * |\{e = (w, v) \in E\}| * \frac{LP(w)}{deg^+(w)} \qquad (5.6)$$

Note that Formula (5.6) only deals with equally weighted relationships. To finally support relationships differently weighted, we must extend it with the average weighting function avg_{Δ} (cf. Sect. 5.2.3):

$$LP(v) = (1 - d) + d \sum_{w \in \Gamma^-(v)} avg_{\Delta}(\{e = (w, v) \in E\}) * |\{e = (w, v) \in E\}| * \frac{LP(w)}{deg^+(w)} \qquad (5.7)$$

Based on Formula (5.7) it becomes possible to determine the link popularity of information objects in a SIN. Note that this corresponds to the solution of a system of equations. Our approach uses an approximate, iterative calculation of the link popularity, i.e. it assigns an initial $LP(v) = init$ to each information object v. The link popularity $LP(v)$ is then iteratively determined for each information object v as follows (let i be the number of iterations). Algorithm 1 shows how the link popularity value for each information object v is calculated[1]:

In summary, Algorithm 1 allows determining the link popularity of information objects based on the relationship structure of a SIN in an iterative way.

---

[1]Our implementation of the SIN Link Popularity Algorithm can be found at http://sourceforge.net/projects/linkinganalyzer/.

Input: $SIN = (V, E, L, W, f_l, f_w); d; i; init;$
Result: $LP(v)$ for each $v \in V(SIN);$
foreach $v \in V(SIN)$ **do** $LP(v) = init$ **foreach** $e \in E(SIN)$ **do** $f_s(e)$ **for** $j = 1$ **to** i **do**
    foreach $v \in V(SIN)$ **do**
        $pop = 0;$
        foreach $w \in \Gamma^-(v)$ **do**
            $pop \overset{+}{=} avg_\Delta(\{e = (w, v) \in E\}) *$
                $|\{e = (w, v) \in E\}| * LP(w) / deg^+(w);$
        end
        $LP(v) = (1 - d) + d * pop;$
    end
    $j = j + 1;$
end

Algorithm 1: SIN Link Popularity Algorithm.

Determining Rate Popularity In the following, we introduce an algorithm that determines the rate popularity of process information based on user ratings. In enterprises, existing IL solutions often allow users to rate the quality of process information, e.g. by means of "like buttons" or "five stars ratings". The set of ratings R can then be used to determine the *rate popularity* $RP(v)$ of an information object v. However, ranking information objects based on user ratings is a non-trivial task. Again, we first show that existing algorithms are not sufficient in POIL and then introduce the *SIN RP algorithm* for determining the rate popularity of information objects (cf. Fig. 5.10).

An approach for determining the rate popularity $R(v)$ of an information object v is to rank information objects by their *total number* of ratings $|R(v)|$:

$$RP(v) = |R(v)| \tag{5.8}$$

Another approach is to determine the rate popularity $RP(v)$ based on the *average user rating* using $avg(R(v))$ of an information object v:

$$RP(v) = \sum_{r \in R(v)} \frac{r}{|R(v)|} \tag{5.9}$$

However, applying Formulas (5.8) or (5.9) is not appropriate in a SIN. Both tend to prefer information objects available for a longer time (i.e. there was more time for users to rate for these information objects). This shortcoming is rather problematic in enterprise environments with continuously emerging information objects. Using Formula (5.9) results in another problem: Assume that in a "five stars rating" there

Fig. 5.10 Rate popularity algorithms

is an information object with an overall weight of 4.8, which is based on hundreds of individual ratings. Additionally assume that another information object is rated by one knowledge worker with 5.0. The latter information object is then directly ranked on the first position. To avoid this, all ratings must be taken into account.

Thus, we calculate the rate popularity with *Bayesian interpretation* (to evaluate the probability of the hypothesis) [33]. Formula (5.10) calculates the average rating $avg(R)$ of information objects. Formula (5.11) then calculates the rate popularity $RP(v)$ of a single information object v taking both the set of ratings R and the age if information objects into account. Thus, we ensure that information objects with few, but favorable ratings are not ranked on the first positions:

$$avg(R) = \sum_{v \in V} \frac{|R(v)| * avg(R(v))}{|R|} \tag{5.10}$$

$$RP(v) = \frac{\left(\frac{|R|}{|\{v \in V | R(v) > 0\}|} * avg(R)\right) + \left(|R(v)| * avg(R(v))\right)}{\frac{|R|}{|\{v \in V | R(v) > 0\}|} + |R(v)|}}{age(v)} \tag{5.11}$$

Algorithm 2 shows how the rate popularity value for each information object v is calculated taking the set of available user ratings R into account[2]:

Input: $SIN = (V, E, L, W, f_l, f_w)$; R;
Result: $RP(v)$ for each $v \in V(SIN)$ where $|R(v)| > 0$;
foreach $v \in V(SIN)$ **do**
    if $|R(v)| > 0$ **then**
        $avg(R) \stackrel{+}{=} |R(v)| * avg(R(v)) / |R|$;
    end
end
foreach $v \in V(SIN)$ **do**
    if $|R(v)| > 0$ **then**
        $pop = ((|R| / |\{v \in V \mid R(v) > 0\}| * avg(R)) + (|R(v)| * avg(R(v))))$;
        $pop = pop / (|R| / |\{v \in V \mid R(v) > 0\}| + |R(v)|)$; $RP(v) = pop / age(v)$;
    end
end

Algorithm 2: SIN Rate Popularity Algorithm.

Altogether, Algorithm 2 allows determining the rate popularity of information objects based on user ratings in an easy way. Further, its popularity values help to determine the relevance of process information.

---

[2]Our implementation of the SIN Rate Popularity Algorithm can be found at http://sourceforge.net/projects/ratinganalyzer/.

5.3   POIL in Practice

This section presents two use cases of POIL. First, Sect. 5.3.1 discusses the role of POIL as enabler of process navigation and visualization support. Second, Sect. 5.3.2 then introduces iCare, an application providing contextualized medical information in patient treatment.

5.3.1   Use Case 1: Process Navigation and Visualization

In enterprises, large process model repositories have emerged [34]. In general, a process model repository not only comprises process models (i.e. process model collections), but also related process information. To cope with this data load, various services for querying, comparing and handling process models as well as related process information have been proposed [35, 36]. However, more advanced concepts enabling an integrated access to both process models and related process information are still missing [37]. To establish a link between the different artifacts, i.e. to provide an integrated view on business process models and related process information, process portals are used. For example, consider the snapshot of a process portal from the automotive domain as shown in Fig. 5.11.

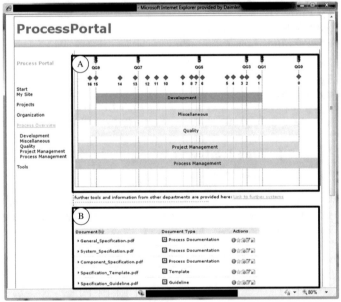

A: Process Model Collection; B: Process Information

Fig. 5.11 Example of a process portal from the automotive domain

Process model collections are visualized in terms of colored rectangles (cf. Fig. 5.11A). The process models of a specific collection may then be accessed by double-clicking on the respective rectangle. Furthermore, process information related to a process model collection or a single process model is presented to users through document lists. In turn, the latter are manually created and maintained by a portal administrator (cf. Fig. 5.11B).

A major drawback of current process portals is that links between process models and process information are defined statically. As shown in [20, 21] this usually leads to rather static process maps [38]. However, users need intuitive support for navigating in large process model collections as well as for accessing related process information in a given context [39]. Thereby, *navigation* refers to the way users may interact with the process model repository. For example, a user may want to navigate from the visualization of an entire process model collection to the one of a single process task enriched with task-specific process information. POIL enables such a flexible navigation as well as various visualizations based on a *navigation space* that is constructed based on the SIN [40].

Running Example For further illustration, we refer to a real-world scenario from the automotive domain. It consists of a process model collection dealing with the development of electric/electronic systems for cars [2]. In detail, the scenario comprises process models related to *requirements engineering*. We consider a *general specification* process (cf. Fig. 5.12) that involves five roles: *E/E Development* (R1), *Component Responsible* (R2), *Expert* (R3), *Project Responsible* (R4), and *Decision Maker* (R5). In addition, the process comprises eleven tasks (i.e. T1–T11) related to the preparation, creation and validation of a general specification of a car component. In turn, these process tasks refer to twelve data objects (D1–D12).

Building a Navigation Space The SIN representation of the process model which results after the initial integration phase is shown in Fig. 5.13. Note that a shared drive is used as a data source for additional information objects which are integrated into the SIN based on semantic and syntactic analyses. For the sake of simplicity, detailed information on single relationship labels and weights are only illustrated in few examples. Further, note that the SIN from Fig. 5.13 is simplified regarding its overall size, i.e. it only covers a part of the actual SIN representing the scenario of the running example. In particular, a SIN may comprise hundreds or thousands of linked process and information objects.

In order to construct the navigation space, first of all, we reorganize the SIN. Specifically, we transform the SIN into a hierarchical structure that allows us to derive three *navigation dimensions*: (1) *semantic* dimension, (2) *geographic* dimension, and (3) *view* dimension. The *semantic dimension* allows displaying process and information objects on different levels of detail. The latter range from abstract process information (e.g. names of process tasks) to a more detailed one (e.g. all information available for process tasks). The *geographic dimension* enables visual zooming without changing the level of detail. Think of a magnifier while reading a newspaper. Finally, the *view dimension* allows users to focus on specific process aspects while eliminating others. For example, a temporal view on a process shall

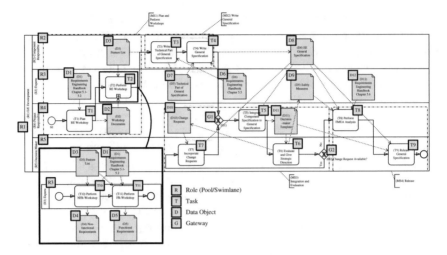

Fig. 5.12 The *general specification* process

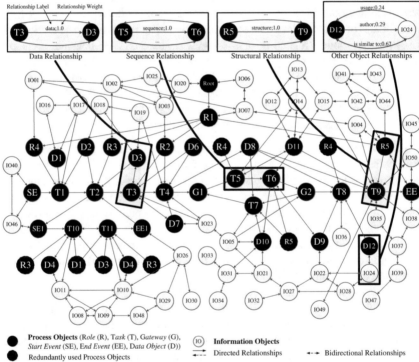

Fig. 5.13 SIN representing the *general specification* process from Fig. 5.12

enable process participants to easily identify deadlines or other temporal constraints [41], whereas an organizational view should provide access to information objects like contact persons or documents [42]. Altogether, these three navigation dimensions form the navigation space. The navigation space is constructed in two steps taking a SIN as input:

Step 1 (Process Space): First, the *process space* is constructed. It represents a harmonized, but preliminary data structure that is used to construct the navigation space. For deriving the process space of a SIN, we first extract the objects related to single process models from the SIN (cf. Fig. 5.14, Step 1.1).

Each extracted process model is then represented as a tree-like structure. This structure is determined by means of specific *structural relationships*, representing hierarchical associations between objects of the SIN. Then, we compose the extracted models to a single structure representing the entire process model collection (cf. Fig. 5.14, Step 1.2). Finally, we add the information objects retrieved from the SIN by following the object relationships between the extracted process objects and their related information objects (cf. Fig. 5.14, Step 1.3).

Step 2 (Navigation Space): Taking the process space derived in Step 1 as input, the *navigation space* can be constructed. In particular, the aforementioned navigation dimensions need to be covered. First, the semantic dimension is constructed based on the tree *levels* of the process space. Thereby, all process and information objects belonging to the same level constitute a particular navigation state (cf. Fig. 5.15, Step 2.1). Second, the geographic dimension extends the semantic one by adding zooming functions (cf. Fig. 5.15, Step 2.2). Third, the view dimension visualizes process and information objects of both the semantic and geographic dimension (cf. Fig. 5.15, Step 2.3). By combining the three navigation dimensions, we obtain the overall navigation space.

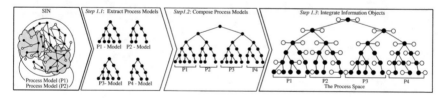

Fig. 5.14 Constructing the process space (Step 1)

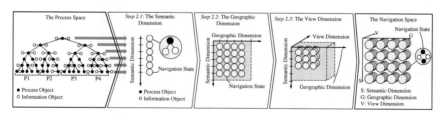

Fig. 5.15 Constructing the navigation space (Step 2)

More detailed information on the construction and formalization of the navigation space can be found in [39, 43].

Implementing the Navigation Space The presented navigation space was implemented in Compass, a tool that aims to support knowledge workers who involved in engineering of electric/electronic components for cars, trucks, and buses. The Compass user interface comprises three major components (cf. Fig. 5.16): First, the *process management area* (cf. Fig. 5.16A) provides management functions (e.g. a breadcrumb navigation and a timeline showing important dates). Second, the *tool area* (cf. Fig. 5.16B) provides functions for interacting with process model collections. Third, contents (i.e. process models and process information) are depicted in the *content area* (cf. Fig. 5.16C).

Compass allows integrating process models with relevant process information. Further, it supports interactions between users and process models. In the latter context, the tool provides three different views: a *time-based view* (cf. Fig. 5.16), a *logic-based view* (cf. Fig. 5.17), and a *text-based view* (not shown). It also implements the presented navigation dimensions (i.e. semantic, geographic, and view dimension). Finally, Compass enables an integrated access to process models and enterprise process information through a single user interface.

Figures 5.16 and 5.17 refer to the running example. Figure 5.16 shows a visualization of the *Requirements Engineering* process model collection. It comprises three process topics (*Component, System,* and *General Specification*) represented as rectangles in the time-based view. Thereby, different colors indicate different roles involved in these process topics. Increasing the geographic and semantic dimension, in combination with a change of the view dimension, allows the user to display the

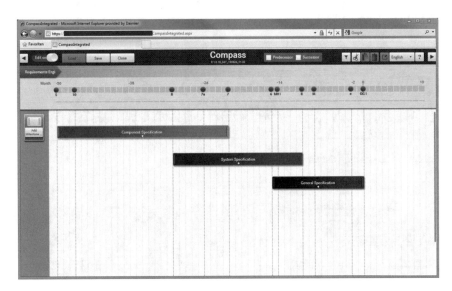

Fig. 5.16 Time-based view on the process model collection

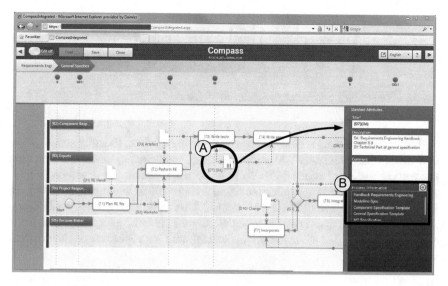

Fig. 5.17 Logic-based view on the *general specification* process

underlying process model (cf. Fig. 5.17), i.e. the logic-based view of the *general specification* process from Fig. 5.12, together with related process information. In this context, data objects $D6$ and $D7$ are displayed as icons (cf. Fig. 5.17A). By clicking on one of these icons, a window on the right hand side is displayed, which provides detailed information about the data object, including a list with related process information (e.g. additional documents such as guidelines or best practices) (cf. Fig. 5.17B).

Compass is currently run by 4 business units of an automotive manufacturer. 364 employees use it during their daily work. Process model collections maintained by Compass comprise between 4 and 50 process models (including between 8 and 37 process tasks) depending on the business unit. 390 documents such as guidelines, checklists and handbooks, are included.

5.3.2 Use Case 2: Medical Information Logistics

The diversity and quantity of medical information emerging in patient treatment and administration makes it a challenging task for medical staff, such as doctors and nurses, to identify and handle the medical information they need to perform their tasks [20]. During a ward round, for example, doctors not only have to rely on patient records, but also on laboratory reports and medical knowledge [44]. Generally, the effective and efficient delivery of medical information is a prerequisite to provide evidence-based decisions, diagnoses and treatments.

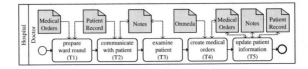

Fig. 5.18 The doctor's ward round (simplified BPMN model)

Today, medical staff is confronted with very limited time. Existing studies show that doctors can spend only 7.5 min for searching and handling medical information per patient and ward round [45]. To improve the situation we developed iCare[3] which is a web-based Java application based on semantic technology [46]. Its overall goal is the personalized delivery of medical information to medical staff. Using iCare, medical staff does not need to search for medical information anymore, but is automatically supplied with relevant medical information dependent on their current work context.

Application Scenario The iCare application scenario has been developed based on the results of an exploratory case study we performed in a university hospital [20, 21]. The focus of this study was the analysis of an unplanned, stationary hospitalization, including patient admission, medical indication in the anesthesia, surgical intervention, post-surgery treatment, patient discharge, and financial accounting and management.

Specifically, iCare aims to support ward rounds (cf. Fig. 5.18). First, the ward round is prepared, i.e. the doctor looks at patient information (e.g. name, pre-existing diseases) and medical orders (e.g. prescribed drugs, current therapy) (task T1). Then, the doctor communicates with the patient and asks for additional information about his health status (task T2). This information is documented. Afterwards, the patient is examined (task T3) and patient information (e.g. pulse rate) is updated accordingly. Finally, the doctor reflects the patient's status and, depending on his assumptions, makes medical orders (e.g. on the procurement of drugs) (task T4). Again, patient information and medical orders are updated accordingly (task T5). Though this process may vary across different hospitals and even within one hospital, it can be found in every hospital.

For each of the tasks shown in Fig. 5.18, a variety of heterogeneous medical information is needed, e.g. patient records, notes, medical orders, laboratory reports, medical guidelines, and patient protocols. This medical information is typically stored in widespread sources like, for example, hospital information systems, medical databases, applications, and libraries. iCare collects and integrates such distributed information if it is electronically available. The home screen of iCare (cf. Fig. 5.19) shows the tasks as introduced above.

---

[3]A screencast presenting the iCare application is available at http://nipro.hs-weingarten.de/screencast.

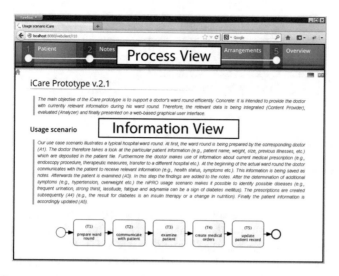

Fig. 5.19 Home screen of iCare

ICare The user interface of iCare is divided into two parts: the *process view* and the *information view*. The former illustrates the currently executed process (i.e. the doctor's ward round), whereas the latter shows the corresponding medical information (e.g. patient records, laboratory reports, medical orders, and notes). iCare works both in desktop browsers and on mobile devices.

The main features of iCare are the *integration* and *analysis* of medical information as well as the *delivery* of needed medical information to medical staff.

- iCare enables the integration of structured, semi-structured, and unstructured medical information electronically available from heterogeneous data sources.
- iCare enables the automatic syntactic and semantic analysis of medical information to determine semantic relationships based on which medical staff can derive and generate new medical knowledge.
- iCare enables the delivery of needed medical information to medical staff, i.e. iCare represents a central access point and unified view on information.

Architecture iCare implements four architectural layers: a data layer, a semantic integration layer, a context layer, and an application layer (cf. Fig. 5.20).

The *data layer* concerns the data sources to be integrated (e.g. hospital information systems, databases, digital libraries, health records etc.). For each data source, a *ContentProvider*[4] is implemented. Its main task is to transform proprietary medical information into a generic, uniform information format. This is a necessary prerequisite for the subsequent syntactic and semantic analysis.

---

[4]These ContentProviders are available as open-source plugins at http://sourceforge.net/directory/?q=iqser.

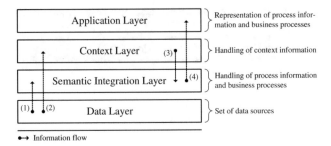

Fig. 5.20 Architecture of iCare

The *semantic integration layer*, in turn, is responsible for the syntactic and semantic analysis of medical information. For this purpose, we use the semantic middleware iQser GIN server [28]. Syntactic and semantic analysis is performed in several steps. In a first step, basic attributes of integrated information such as authorships are compared (~ *syntactic analysis*). This allows, for example, linking information with the same author (e.g. a specific doctor). Second, the raw full text of all available information is analyzed (~ *semantic analysis*). For this purpose, algorithms from the fields of data mining, text mining (e.g. text preprocessing, linguistic preprocessing, clustering, classification, information extraction), pattern-matching, and machine learning (e.g. supervised learning, unsupervised learning, reinforcement learning, transduction) are applied [23]. The goal is to further classify and group correlated information. Finally, user behavior is investigated, for example, the frequency of using certain information in the context of specific process tasks. The result of the analysis is a SIN. In particular, the SIN allows identifying information linked to each other in the one or other way, e.g. information addressing the same topic (e.g. "flu") or needed when performing a particular process task (e.g. "prepare ward round") [26].

The *context layer* is responsible for integrating and analyzing context information (e.g. used device, location, time, user behavior). In [10], we have described a framework realizing the context layer. Context information is gathered from data sources called sensors. We distinguish between physical sensors (e.g. thermometer), virtual sensors (e.g. keyboard input), and logical sensors (e.g. sensors which allow to detect a doctor's position by analyzing logins at devices and a mapping to locations). In addition, further context information can be also derived from existing ones (e.g. by aggregation or reduction). A CM, which is constructed based on available context information, allows characterizing a doctor's work context which can then be used to filter the SIN. Note that the CM is completely independent from the SIN, i.e. context information is only stored in the CM but not in the SIN (see [10] for details).

Finally, the *application layer* concerns the personalized delivery of medical information. The application layer is responsible for the joint presentation of executed processes (or tasks) and corresponding medical information. Further details regarding the layers can be found in [19].

Using iCare In the following we describe how our scenario can be supported by iCare.[5] To support task T1, a search box is offered to select single patients. After having selected a patient, iCare provides available information such as name, pre-existing diseases, gender, weight, and date of birth, from the respective patient record. When performing task T2, existing medical notes for the previously selected patient are shown, i.e. information about the patient's health status. Upon need, the doctor can add, update, or delete medical notes. Based on an analysis of available medical information, potential diseases and treatment options are then automatically determined when performing task T3. For example, the analysis takes into account the patient record, medical notes, and medical information from Onmeda and can automatically conclude that sore throat, croakiness, rheumatic pains and absence of appetite are potentially caused by the disease "flu". As an additional result of the syntactic and semantic analysis, the doctor is also supplied with treatment options which are also automatically determined. If a treatment option is selected, a more detailed treatment description and respective instructions are displayed. In task T4, the doctor can then add or update medical orders. Finally, the patient record, medical notes, and medical orders can be updated in task T5.

In summary, iCare supports the doctor's ward round by reducing the time for searching and handling medical information. iCare automatically delivers needed medical information depending on the current work context.

5.4 Discussion

In recent years, various approaches have been proposed to realize selected concepts, including data warehouses, business intelligence, decision support systems, and enterprise content management. However, these approaches have not primarily been designed with POIL and its underlying concepts in mind [19].

Data warehouses rather focus on creating an integrated database [47]. Opposed to this, POIL deals with the management of process information flows to support the execution of business processes. In turn, traditional business intelligence addresses data analytics and is typically completely isolated from business process execution [48]. Moreover, information supply is often restricted to decision makers at the management level [49, 50]. By contrast, POIL focuses on the integration and analysis of process information as well as its delivery to both knowledge workers and decision makers. In turn, decision support systems support decision-making, i.e. they serve the management level [51]. Opposed to this, enterprise content management deals with the management of information across enterprises referring to related strategies, methods and tools [52].

Missing process-awareness in contemporary approaches has guided the development of POIL. Generally, the goal of POIL is to provide the right process information,

---

[5]Since we have no access to international digital medical libraries we use the German health portal Onmeda (http://www.onmeda.de) instead. Therefore, some screenshots contain German text.

Fig. 5.21 Problem
dimensions of POIL

in the right format and quality, at the right place, at the right point in time, and to the right people (both knowledge workers and decision makers). In particular, process participants should not have to actively search for relevant process information anymore, but be automatically supplied with needed process information (even if their work context is dynamically changing).

Unlike previous approaches, POIL combines information-, context-, and process-awareness (cf. Fig. 5.21). POIL is *information-aware* as it allows effectively handle process information. POIL is *context-aware* as it supports the use of context information to characterize the process participant's situation. Finally, POIL is *process-aware* as it allows integrating and analyzing business processes and their tasks (both process schemas and process instances).

5.5 Summary

Enterprises are confronted with an increasing amount of data. A major problem is to align process-related information with business processes. So far, these have usually been handled separately, e.g. through shared drives, databases, and portals on one hand and process management technology on the other.

This chapter suggests a novel approach called *process-oriented information logistics* (POIL) to bridge this gap. Specifically, the contribution of this chapter is twofold: First, it introduced basic POIL concepts. Second, it demonstrated the application and benefits of POIL in practice.

Acknowledgements This work was done in the niPRO research project. The project is funded by the German Federal Ministry of Education and Research (BMBF) under grant number 17102X10. More information can be found at http://www.nipro-project.org.

References

1. Lenz, R., Reichert, M.: IT support for healthcare processes—premises, challenges, perspectives. J. Data Knowl. Eng. **61**(1), 39–58 (2007)
2. Müller, D., Herbst, J., Hammori, M., Reichert, M.: IT support for release management processes in the automotive industry. In: Proceedings of 4th International Conference on Business Process Management (BPM'06), pp. 368–377 (2006)

3. Edmunds, A., Morris, A.: The problem of information overload in business organisations: a review of the literature. Int. J. Inf. Manage. **20**(1), 17–28 (2000)
4. Rowley, J.: The wisdom hierarchy: representations of the dikw hierarchy. J. Inf. Sci. **33**(2), 163–180 (2006)
5. Girit, H., Eberhard, R., Michelberger, B., Mutschler, B.: On the precision of search engines: results from a controlled experiment. In: Proceedings of 5th International Conference on Business Information Systems (BIS'12), pp. 201–212 (2012)
6. Bocij, P., Chaffey, D., Greasley, A., Hickie, S.: Business Information Systems: Technology. Prentice Hall, Development and Management for the E-Business (2006)
7. Michelberger, B., Mutschler, B., Reichert, M.: Towards process-oriented information logistics: why quality dimensions of process information matter. In: Proceedings of 4th International Workshop on Enterprise Modelling and Information Systems Architectures (EMISA'11), pp. 107–120 (2011)
8. Künzle, V., Reichert, M.: PHILharmonicFlows: towards a framework for object-aware process management. J. Softw. Maint. Evol. Res. Pract. **23**(4), 205–244 (2011)
9. Hipp, M., Michelberger, B., Mutschler, B., Reichert, M.: A framework for the intelligent delivery and user-adequate visualization of process information. In: Proceedings of 28th Symposium On Applied Computing (SAC'13), pp. 1383–1390 (2013)
10. Michelberger, B., Mutschler, B., Reichert, M.: A context framework for process-oriented information logistics. In: Proceedings of 15th International Conference on Business Information Systems (BIS'12), pp. 260–271 (2012)
11. Haseloff, S.: Context awareness in information logistics. PhD Thesis, Technical University of Berlin (2005)
12. Gronau, N., Müller, C., Korf, R.: KMDL—capturing, analysing and improving knowledge-intensive business processes. J. Univ. Comput. Sci. (JUCS) **11**(4), 452–472 (2005)
13. Heuwinkel, K., Deiters, W.: Information logistics, E-Healthcare and trust. In: Proceedings of International Conference on e-Society (IADIS'03), pp. 791–794 (2003)
14. Deiters, W., Löffeler, T., Pfennigschmidt, S.: The information logistics approach toward user demand-driven information supply. In: Proceedings of Conference on Cross-Media Service Delivery (CMSD'03), pp. 37–48 (2003)
15. Womack, J.P., Jones, D.T.: Lean Thinking: Banish Waste and Create Wealth in Your Corporation. Free Press (2003)
16. Michelberger, B., Andris, R., Girit, H., Mutschler, B.: A literature survey on information logistics. In: Proceedings of 16th Internationl Conference on Business Information Systems (BIS'13), pp. 138–150 (2013)
17. Dinter, B., Winter, R.: Information logistics strategy—analysis of current practices and proposal of a framework. In: Proceedings of 42nd Hawaii International Conference on System Sciences (HICSS'09), pp. 1–10 (2009)
18. Winter, R.: Enterprise-wide information logistics: conceptual foundations, technology enablers, and management challenges. In: Proceedings of 30th International Conference on Information Technology Interfaces (ITI'08), pp. 41–50 (2008)
19. Michelberger, B., Mutschler, B., Reichert, M.: Process-oriented Information logistics: aligning enterprise information with business processes. In: Proceedings of 16th IEEE International Enterprise Computing Conf (EDOC'12), pp. 21–30 (2012)
20. Michelberger, B., Mutschler, B., Reichert, M.: On handling process information: results from case studies and a survey. In: Proceedings of 2nd International Workshop on Empirical Research in Business Process Management (ER-BPM'11), pp. 333–344 (2011)
21. Hipp, M., Mutschler, B., Reichert, M.: On the context-aware, personalized delivery of process information: viewpoints, problems, and requirements. In: Proceedings of 6th International Conference on Availability, Reliability and Security (ARES'11), pp. 390–397 (2011)
22. Rinderle, S., Reichert, M., Dadam, P.: On dealing with structural conflicts between process type and instance changes. In: Proceedings of 2nd International Conference on Business Process Management (BPM'04), pp. 274–289 (2004)

23. Wurzer, J.: New approach for semantic web by automatic semantics. In: Proceedings of 2nd European Conf on Semantic Technology (ESCT'08) (2008)
24. Hotho, A., Nürnberger, A., Paaß, G.: A brief survey of text mining. J. Comput. Linguisti. Lang.Technol. **20**(1), 19–62 (2005)
25. Michelberger, B., Ulmschneider, K., Glimm, B., Mutschler, B., Reichert, M.: Maintaining semantic networks: challenges and algorithms. In: Proceedings of 16th International Conference on Information Integration and Web-based Applications & Services (iiWAS'14), pp. 365–374 (2014)
26. Michelberger, B., Mutschler, B., Hipp, M., Reichert, M.: Determining the link and rate popularity of enterprise process information. In: Proceedings of 21st International Conference on Cooperative Information Systems (CoopIS'13), pp. 112–129 (2013)
27. Michelberger, B., Mutschler, B., Binder, D., Meurer, J., Hipp, M.: iGraph: intelligent enterprise information logistics. In: Proceedings of 10th International Conference on Semantic Systems (SEMANTiCS'14), Posters & Demonstrations Track, pp. 27–30 (2012)
28. Wurzer, J., Mutschler, B.: Bringing innovative semantic technology to practice: the iqser approach and its use cases. In: Proceedings of 4th International Workshop on Applications of Semantic Technologies (AST'09), pp. 3026–3040 (2009)
29. Diestel, R.: Graph Theory. Springer (2010)
30. Gyöngyi, Z., Garcia-Molina, H., Pedersen, J.: Combating web spam with TrustRank. In: Proceedings of 13th International Conference on Very Large Data Bases (VLDB'04), 30, pp. 576–587 (2004)
31. Borodin, A., Roberts, G.O., Rosenthal, J.S., Tsaparas, P.: Link analysis ranking: algorithms, theory, and experiments. J. ACM Trans. Internet Technol. (TOIT) **5**(1), 231–297 (2005)
32. Page, L., Brin, S., Motwani, R., Winograd, T.: The PageRank Citation Ranking: Bringing Order to the Web. Technical Report, Stanford University (1999)
33. MacKay, D.J.C.: Information Theory. Cambridge University Press, Inference and Learning Algorithms (2003)
34. Weber, B., Reichert, M., Mendling, J., Reijers, H.A.: Refactoring large process model repositories. J. Comput. Ind. **62**(5), 467–486 (2011)
35. Rosa, M.L., Reijers, H.A., van der Aalst, W.M.P., Dijkman, R.M., Mendling, J., Dumas, M., García-Bañuelos, L.: APROMORE: an advanced process model repository. J. Expert Syst. Appl. **38**(6), 7029–7040 (2011)
36. Kunze, M., Weske, M.: Metric trees for efficient similarity search in process model repositories. In: Proceedings of 1st International Workshop Process in the Large (IW-PL'10), pp. 535–546 (2010)
37. van der Aalst, W.M.P.: Process-aware information systems: design, enactment and analysis. In: Wiley Encyclopedia of Computer Science and Engineering, pp. 2221–2233 (2009)
38. van Wijk, J.J., Nuij, W.A.A.: A model for smooth viewing and navigation of large 2D information spaces. J. IEEE Trans. Visual. Comput. Graphics **10**(4), 447–458 (2004)
39. Hipp, M., Michelberger, B., Mutschler, B., Reichert, M.: Navigating in process model repositories and enterprise process information. In: Proceedings of IEEE 8th International Conference on Research Challenges in Information Science (RCIS'14), pp. 1–12 (2014)
40. Hipp, M., Strauss, A., Michelberger, B., Mutschler, B., Reichert, M.: Enabling a user-friendly visualization of business process models. In: Proceedings of 3rd International Workshop on Theory and Applications of Process Visualization (TaProViz'14), BPM 2014 Workshops (2014) (accepted for publication)
41. Lanz, A., Weber, B., Reichert, M.: Time Patterns for Process-Aware Information Systems. J. Requir. Eng. 113–141 (2014)
42. Pryss, R., Mundbrod, N., Langer, D., Reichert, M.: Modeling the resource perspective of business process compliance rules with the extended compliance rule graph. In: Proceedings of 15th International Working Conference on Business Process Modeling, Development, and Support (BPMDS'14), pp. 48–63 (2014)
43. Hipp, M., Mutschler, B., Reichert, M.: Navigating in complex business processes. In: Proceedings of 23rd International Conference on Database and Expert Systems Applications (DEXA'12), pp. 466–480 (2012)

44. Pryss, R., Mundbrod, N., Langer, D., Reichert, M.: Supporting medical ward rounds through mobile task and process management. J. Inf. Syst. e-Bus. Manag. (2014)
45. Weber, H., Stöckli, M., Nübling, M., Langewitz, W.A.: Communication during ward rounds in internal medicine: an analysis of patient-nurse-physician interactions using RIAS. In: Proceedings of European Association for Communication in Healthcare (EACH'07) Conference, 67(3), pp. 343–348 (2007)
46. Michelberger, B., Reisch, A., Mutschler, B., Wurzer, J., Hipp, M., Reichert, M.: iCare: intelligent medical information logistics. In: Proceedings of 15th International Conference on Information Integration and Web-based Applications & Services (iiWAS'13), pp. 396–399 (2013)
47. Lechtenbörger, J.: Data warehouse schema design. Infix Akademische Verlagsgesellschaft Aka GmbH, PhD Thesis, University of Münster (2001)
48. Bucher, T., Dinter, B.: Process orientation of information logistics—an empirical analysis to assess benefits, design factors, and realization approaches. In: Proceedings of 41st Hawaii International Conference on System Sciences (HICSS'08), pp. 392–402 (2008)
49. Baars, H., Kemper, H.G.: Management support with structured and unstructured data—an integrated business intelligence framework. J. Inf. Syst. Manag. 25(2), 132–148 (2008)
50. Rouhani, S., Asgari, S.: Review study: business intelligence concepts and approaches. Am. J. Sci. Res. 50, 62–75 (2012)
51. Janakiraman, V.S., Sarukesi, K.: Decision Support Systems. Prentice Hall (2004)
52. Cameron, S.A.: Enterprise Content Management: A Business and Technical Guide. British Informatics Society (2011)

Chapter 6
A Predictive Approach Enabling Process Execution Recommendations

Johannes Schobel and Manfred Reichert

Abstract In enterprises, decision makers need to continuously monitor business processes to guarantee for a high product and service quality. To accomplish this task, process-related data needs to be retrieved from various information systems—periodically or in real-time—and then be aggregated based on key performance indicators (KPIs). If target values of the defined KPIs are violated (e.g., production takes longer than a predefined threshold), the reasons of these violations need to be identified. In general, such a retrospective analysis of business process data does not always contribute to prevent respective key performance violations. To remedy this drawback, process-aware information systems (PAIS) should enable the automated identification of processes, which are not well performing, and support users in executing these processes through recommendations. For example, it should be indicated, which problems might occur in future when taking the current course of the process instance as well as previous process instances into account. This chapter presents a methodology as well as an architecture for the support of predictive process analyses. In this context, algorithms from machine learning are applied to compare running process instances with historic process data and to identify diverging processes. In particular, the predictive approach will enable enterprises to quickly react to upcoming problems and inefficiencies.

6.1 Introduction

Enterprises are increasingly forced to quickly react to changing customer needs as well as to reduce both, time-to-market and production costs. To meet these demands, *Process-Aware Information Systems* (PAISs) [1] for defining, coordinating, monitoring, and optimizing business processes have been adopted [2].

J. Schobel (✉) · M. Reichert
Institute of Databases and Information Systems, Ulm University, Ulm, Germany
e-mail: johannes.schobel@uni-ulm.de

M. Reichert
e-mail: manfred.reichert@uni-ulm.de

© Springer International Publishing AG 2017
G. Grambow et al. (eds.), *Advances in Intelligent Process-Aware Information Systems*, Intelligent Systems Reference Library 123, DOI 10.1007/978-3-319-52181-7_6

155

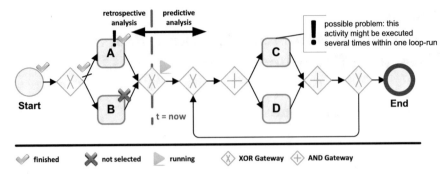

Fig. 6.1 Retrospective and predictive evaluation of a business process instance

To measure and monitor business performance at the operational level, *Business Intelligence* (BI) systems are widely used in practice [3]. In this context, various metrics (e.g., *Key Performance Indicators*, KPIs) are used to assess process performance with respect to the achievement of Pre-specified business goals [4]. Contemporary BI systems periodically extract operational data from information systems, aggregate and Pre-process the data, and then present them to decision makers, e.g., using dashboards [5]. However, periodical data extraction does not allow for the real-time monitoring of business processes and their performance. Accordingly, existing BI systems only enable a retrospective view on completed business process instances (i.e., business cases) [6]. In particular, business process instances not performing well can solely be identified retrospectively (e.g., monthly); i.e., a prospective reaction on upcoming problems is not supported by existing BI systems (cf. Fig. 6.1).

Many BI systems lack a process-centric view, which is indispensable when targeting at the agile enterprise being able to quickly react to emerging problems and environmental changes [7]. To close this gap, *Business Process Intelligence* (BPI) has emerged as a new discipline [8]. Amongst others, BPI allows visualizing KPIs with respect to business process models, that, in turn, are derived based on process mining techniques [9]. Furthermore, BPI provides advanced mechanisms for a (near) real-time data extraction. Still, the drawback of only having a retrospective view on business process performance remains.

For decision makers it is of utmost importance to detect potential problems at the operational level as soon as possible—best case, respective problems can be predicted before they occur. In this context, the goal of *Predictive Business Process Intelligence* is to enable predictive process analyses, i.e., to assess the future course of a running business process instance as well as to properly react to potential problems emerging in future (cf. Fig. 6.1).

This chapter is organized as follows: Sect. 6.2 introduces background information. Section 6.3 presents a methodology for *Predictive BPI* as well as a related architecture. Section 6.4 introduces an algorithm based on machine learning concepts; in particular, the latter are customized to be applicable to process execution data (i.e., event logs). In addition, recommendations are provided to decision makers in order to ensure proper actions for process instances being in trouble. Section 6.5 presents evaluation results we obtained when applying the approach to execution data from real-world business processes. Section 6.6 discusses related work and Sect. 6.7 concludes the chapter.

6.2   Background

This section presents basic concepts needed for understanding this chapter. First of all, a *process-aware information system* corresponds to an information system enabling support for the enactment of business processes. On one hand, this includes support for the design of business processes and, on the other, it covers business process execution. In modern PAISs, business process logic is strictly separated from application code. Each business process to be executed by the PAIS, therefore, needs to be formally specified in terms of a *process model* (cf. Fig. 6.1). A process model, in turn, correlates to a directed graph with one start- and one end-point. Furthermore, it comprises process steps (e.g., the activities to be executed) as well as control connectors defining their execution order. In addition, gateways (e.g., AND or XOR split gateways) may be used in order to enhance the business process with the respective logic (e.g., certain activities may be executed in parallel or alternatively to each other). Besides the *control flow*, the flow of *data* within a business process may be considered as well. In particular, one may specify *data elements* that are connected with activities via data connectors. Those *connectors*, in turn, allow reading data from or writing data to respective data elements (i.e., by using READ or WRITE connectors).

Executable process models are deployed to *process engines*, which support the creation of corresponding *process instances* and their execution according to the defined process logic. During process execution, a process engine logs run-time data, like the point in time an activity is started, the duration needed to finish a task, or the data produced during its execution.

6.3   Business Process Intelligence

Section 6.3.1 introduces a BPI methodology along the data life cycle (from *Data Import* to *Process Control*). Based on it, Sect. 6.3.2 presents a corresponding BPI system architecture.

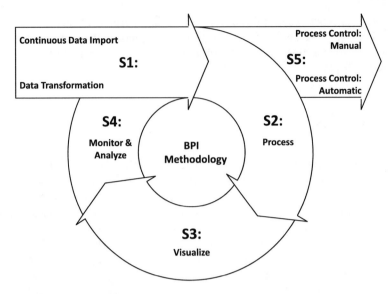

Fig. 6.2 BPI methodology

6.3.1 Methodology

The BPI methodology (cf. Fig. 6.2) was derived by analyzing the requirements of real-world applications for BPI systems. In step *S1*, process execution data provided by the various information systems (e.g., ERP/CRM systems) is extracted, transformed and loaded into the BPI system [10]. In addition, the gathered data is correlated with the respective process model. Step *S2* then processes the extracted data (i.e., data mining, KPI calculation), whereas step *S3* visualizes the corresponding results (e.g., using dashboards and diagrams). These diagrams are then analyzed and continuously monitored by the BPI system as well as decision makers in step *S4*. Based upon the insights gained, the BPI system recommends proper actions to decision makers with the goal to improve overall process performance. In order to apply these recommendations, step *S5* allows for adapting already running process instances.

6.3.2 Architecture

Reviewing state-of-the-art BI systems and refining the presented BPI methodology, we derive a general architecture for BPI systems (cf. Fig. 6.3) that comprises the following components:

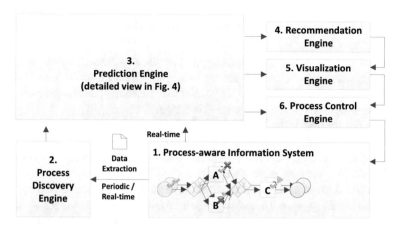

Fig. 6.3 BPI system architecture

1. **Process-Aware Information System (PAIS)**: A PAIS executes business processes and records related events in respective execution logs (e.g., *activity started*, *process instance aborted*).
2. **Process Discovery Engine**: If no explicit process model is available, the latter can be discovered through mining of execution logs. This is required in the considered approach as the algorithms provided by the *Prediction Engine* (cf. Sect. 6.4) presume the presence of a process model. In addition, the derived process model needs to be enriched with process data elements (e.g., customer name).
3. **Prediction Engine**: Process models and related execution data are analyzed with the *Prediction Engine*. Data from already completed process instances is separated and used as input for unsupervised learning techniques (cf. Sect. 6.4.1). Results are then applied to the not yet completed process instances with the goal to predict performance metrics or malfunctions.
4. **Recommendation Engine**: Information about the execution of a process instance as well as results of the *Prediction Engine* are merged. To properly assist decision makers, *recommendations* for optimizing the continuation of already running process instances are derived, e.g., to execute activity A before B in order to avoid unnecessary waiting time. For this purpose, historic process data is considered (cf. Sect. 6.4.2).
5. **Visualization Engine**: Analysis results as well as the determined recommendations are visualized, e.g., by using dashboards. Based on these visualizations, decision makers decide on the further course of executing the process instance.
6. **Process Engine**: To avoid potential problems upcoming during process execution, optimizations detected by the *Prediction Engine* (e.g., to change staff assignment of an activity) must be propagated back to the PAIS. This is accomplished by a Process Control Engine.

This system architecture supports the described BPI methodology in a proper way.

6.4 Predictive Business Process Intelligence

Section 6.4 presents basic algorithms applied by the *Prediction Engine*. More precisely, algorithms from neuro science and bioinformatics are applied in combination with key performance indicators (KPIs). Such a "mashup" enables predictions on the further progress of a running process instance.

Figure 6.4 illustrates the architecture of the *Prediction Engine*, which relies on the *Execution Repository* to store the execution data extracted from running as well as completed process instances. In particular, this repository serves as data source for *unsupervised learning algorithms* [11].

Section 6.4.1 presents a customized version of the *Edit Distance algorithm*, which calculates the *edit distances* between process instances, i.e., the number of changes required to convert one process instance into an other. Note that this constitutes a prerequisite for clustering similar process instances. Section 6.4.2 then shows how the calculated distances can be used to extract recommendations on further actions concerning not well-performing and still running process instances.

6.4.1 Predictive Process Analysis

We present the approach for *predictive process analysis* of a given process instance.

Definition 1 (*Dimension of a Process Instance*)

Let $i = \langle T, D_{input}, D_{output} \rangle$ be a *process instance*, derived from a particular process model, and consisting of an execution log T as well as input and output data element values (i.e., D_{input} and D_{output}). We denote T as well as all elements from D_{input} and D_{output} as *dimensions* of the process instance i.

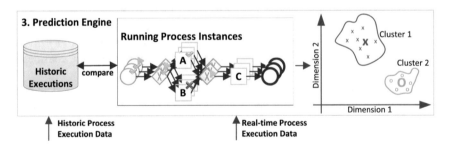

Fig. 6.4 Detailed view on the prediction engine

Consider the following example:

$$T = \langle A, B, C, D, E \rangle$$
$$D_{input} = \langle (effort = \{10, 20\}), urgency = \{100\}) \rangle$$
$$D_{output} = \langle \rangle$$

Note, that each of these sets has a different *size* with respect to the dimensions. For example, $size(T) = 5$; thereby, each element of T may comprise of additional elements (e.g., the performing agent of a node or the duration of the execution). Definition 2 defines the sets required by the *Prediction Engine*.

Definition 2 (*Clusters*) Let I be the set of all process instances corresponding to a given business process model and executed by a process-aware information system. Then:

- For any process instance i, $status(i)$ returns its current execution status (i.e., $status(i) \in \{completed, aborted, running\}$).
- $I_{completed} \equiv \{i \in I \mid status(i) \in \{completed, aborted\}\}$ and $I_{running} \equiv \{i \in I \mid status(i) \in \{running\}\}$.
- $I_{training}$ represents the training set with $I_{training} \subset I_{completed}$; furthermore, the test set i_{test} is defined as $I_{test} = I_{completed} \setminus I_{training}$.

In the context of the following analysis, the *(fuzzy) k-means algorithm* [12] is applied to cluster process instances based on their dimensions.

To calculate the similarity between two process instances, the *Edit Distance* (or Levenshtein Distance) algorithm is applied [13]. The Edit Distance is a metric to determine the minimum number of change operations (i.e., insert, update, delete) needed to convert a sequence of symbols into another one. In the approach presented in this chapter we customize this algorithm applying it to execution data of respective process instances. More precisely, each process instance from set $I_{running}$ is compared with each completed process instance of set $I_{completed}$. As instances of set $I_{running}$ have not been completed yet, vectors of different length have to be compared (e.g., some activities may not have been executed yet). Obviously, this can be achieved by cutting off the entries for the not yet executed activities of the running process instances.

Note that in Algorithm 1, parameters $i_r \in I_{running}$ and $i_c \in I_{completed}$ (Line 1) represent the instances to be compared. In turn C_{edit}, C_{new} and C_{del} (input parameters) indicate weights of the operations required to transform one instance into another one (i.e., the insert, update or delete an activity). The algorithm first fills the *distance matrix* D, which represents the the number of operations to transform one instance into another, with zeros (Line 2). Then, the first row and column of every dimension is filled with the associated costs C_{del} and C_{new} (Lines 3–8). Following this, for every cell of the multi-dimensional matrix, the minimum cost is calculated (Lines 9–24). The multiplicative factor $Dimension\ Cost(k)$, shown in Line 21, corresponds to the weight of the currently calculated dimension (e.g., the actor performing an activity

Algorithm 1: Calculating the Edit Distances

Data:
i_r: A process instance which is running
i_c: An already completed process instance
C_{edit}: The costs to edit one activity
C_{new}: The costs to insert a new activity
C_{del}: The costs to delete an activity
$DimensionCost$: The costs for each dimension
Result:
D: The calculated distance matrix for the instances

```
 1  begin
        /* initialize an "empty" distance matrix                        */
 2      D = zeros(size(i_r), size(i_c), size(i_c.dim))
        /* set the first row and column with initial values             */
 3      for (i = 1:size(i_r)) do
 4      |   D(i+1,1,:) = D(i,1) + C_del
 5      end
 6      for (j = 1:size(i_c)) do
 7      |   D(1,j+1,:) = D(1,j) + C_new
 8      end
        /* for each dimension in each instance do                       */
 9      for (k = 1:size(i_r.dim)) do
10      |   for (i = 1:size(i_r)) do
11      |   |   for (j = 1:size(i_c)) do
                    /* are we comparing the same activities?            */
12      |   |   |   if (i_r(k, i) == i_c(k, j)) then
13      |   |   |   |   EditCost = 0
14      |   |   |   else
15      |   |   |   |   EditCost = C_edit
16      |   |   |   end
17      |   |   |   D(i+1,j+1,k) = min(
18      |   |   |       D(i,j,k) + EditCost,
19      |   |   |       D(i+1,j,k) + C_del,
20      |   |   |       D(i,j+1,k) + C_new
21      |   |   |   ) * DimensionCost(k)
22      |   |   end
23      |   end
24      end
25      return D
26  end
```

---

might not be as important as the control flow of the process, as other actors may process this activity as well).

As an example consider the execution of two instances $i_{c1} = \langle A, B, C, D, E \rangle$ and $i_{r1} = \langle A, A, B, E \rangle$. We calculate the distance between them. Recall, that the distance is defined as the number of change operations (i.e., insert, update, delete activities) needed to convert a sequence of symbols into another one. In this case, the distance between i_{c1} and i_{r1} corresponds to 3 (cf. Fig. 6.5).

The framed values shown in Fig. 6.5 which are connected by lines, indicate the *shortest path*—regarding the number of change operations (i.e., insert, update, delete)—to transform process instance i_{r1} into i_{c1}. Thereby, the minimum distance between two instances is always located in the bottom right corner of the matrix. Moreover, lower distance correlates with a higher similarity of the instances. Note that most likely, this indicates a higher chance for similar execution regarding the further course of action.

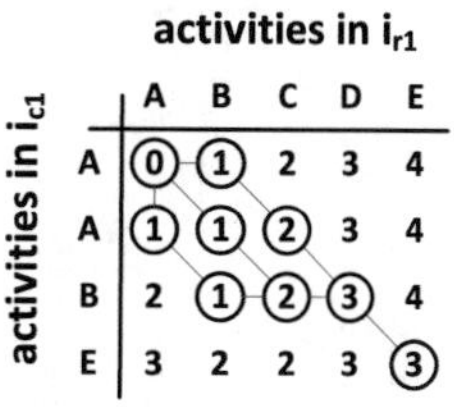

Fig. 6.5 Edit distance
matrix of the dimension
"Control flow"

6.4.2   Recommendations on Process Instances

We apply the distances to derive recommendations for decision makers. For this purpose, we define $i_{closest} \in I_{completed}$ as the *completed process instance being most similar* to a running instance $i_r \in I_{running}$ and $i_c \in I_{completed}$ with

$$EditDist(i_r, i_{closest}, C_{edit}, C_{new}, C_{del}) \leq EditDist(i_r, i_c, C_{edit}, C_{new}, C_{del})$$

In addition, the last element of the execution log T of i_r (cf. Definition 1) in $i_{closest}$ is determined. Then, all dimensions can be mapped from $I_{closest}$ to the considered process instance i_r. Based on this, *estimated* values (e.g., throughput time, costs) may be calculated and derived. Thereby, the distance between i_r and $i_{closest}$ serves as an element of uncertainty for the calculated dimensions. Thereby, a lower distance (i.e., a higher similarity between the process instances) indicates a higher precision with respect to the calculated values. With increasing progress of the running instance, uncertainty with respect to future problems decreases as well.

In summary, this section presented an approach for predictive BPI. It first presented an algorithm to calculate the distance between two process instances. In this context, we adopted the modified *Edit Distance* algorithm and applied it to process instances along different dimensions. Then it was shown, how the prediction approach can be used to provide recommendations to decision makers regarding the further execution of a process instance, i.e., to avoid possible problems in the course of the process instance.

6.5   Evaluation

We applied the discussed approach in a case study to evaluate it (cf. Sect. 6.5.2). Before discussing results, the considered process scenario is presented (cf. Sect. 6.5.1).

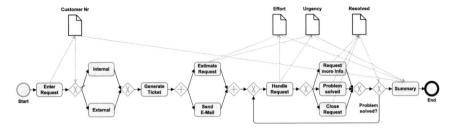

Fig. 6.6 Incident management process (Simplified)

6.5.1 Process Scenario

The evaluation considers an *incident management process* running in a small and medium-sized company in Germany. The process is triggered by a customer when requesting service for a particular product. The PAIS then first checks whether the service request is internal or external, and then creates the ticket referring to this request. Following this, an e-mail is sent to the service staff, which rates the ticket taking the estimated urgency and efforts required to solve it into account. In turn, this triggers the actual handling of the service request, deciding first on how to proceed with it. Either the service request has to be modified (and the applicant be notified accordingly) or the service request is canceled or the ticket is resolved. If the problem still persists after the service staff claims to have resolved it, the handling of the service request restarts. Finally, the service request is logged, billed and archived (cf. Fig. 6.6).

The execution data (i.e., the execution log as well as data element values) are extracted and gathered from the *AristaFlow BPM Suite*, which was used as engine to execute the business process instances [14].

6.5.2 Results

We used a representative subset of 76 *incident management* process instances: 4 of these instances are still running, 8 instances failed, and 64 instances were completed.

In a first step, the provided data (i.e., the execution logs and values of process data elements) are preprocessed. This includes the transformation and anonymization of the data. In a second step, the preprocessed data is loaded into the *Prediction Engine*, which is implemented based on MATLAB [15].

Figure 6.7 shows the results of the (fuzzy) k-means-cluster-algorithm (cf. Sect. 6.4.1) after applying it to the dimensions *performing agents* and *started activities*. It depicts all training instances (i.e., $I_{training}$) as dots ($\bullet$), whereas $C1$, $C2$ and $C3$ represent the cluster centers ($\times$) of the selected dimensions. The thin dashed lines illustrate the test instances (i.e., I_{test}) for these cluster centers, whereas the solid

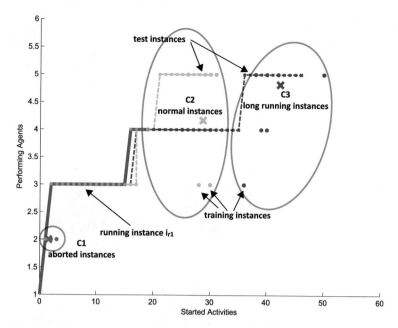

Fig. 6.7 Cluster visualization of process instances

line represents the currently running process instance. In other words, the process instance i_{r1} is executed as expected. Furthermore, no interventions are needed in order to *correct* the process instance.

Moreover, when applying the algorithm to all running and all test instances we obtain different clusters (cf. Sect. 6.4.1). In order to visualize the distances between the instances, we used *heightmaps* (cf. Fig. 6.8). The x-axis (*Running Instances*) lists

Fig. 6.8 Distance with respect to control flow differences for instance i_{r1}

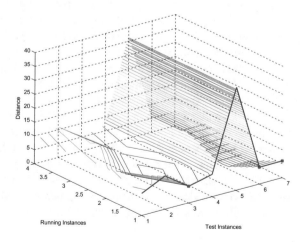

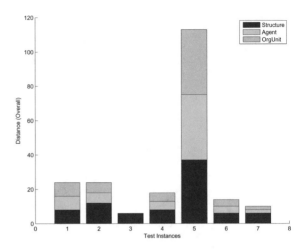

Fig. 6.9 Distances between instance i_{r1} and test instances

all running instances, whereas the z-axis (*Test Instances*) depicts the randomly chosen test instances. The y-axis (*Distance*) indicates the distance between the process instances: i.e., the solid line at x = 1 represents the distance for process instance i_{r1} compared to all chosen test instances—dots on this line show the minimum distance (i.e., highest similarity) to other process instances. For x = 1, for example, the test instances #3, #6 and #7 show the highest similarity to i_{r1}.

In order to obtain a deeper understanding of the behaviour of a particular process instance, all dimensions (i.e., control flow, data element values) are taken into account. Fig. 6.9 depicts the distances between the test instances and the running instance i_{r1}. The bottom block shows the distance with respect to the control flow (cf. Fig. 6.8). For example, one can see, that for these specific instances, test instance #3 is the best matching one (there are only differences regarding control flow), whereas #5 is the least matching one (i.e., large variety in all dimension).

The *Recommendation Engine* (cf. Sect. 6.4.2) suggests continuing with the execution of this process instance according to test instance #3. Note that the different dimensions might influence the result based on their individual weights (e.g., dimension *control flow* is more important than dimension *employee*).

To evaluate the obtained recommendations, we compared the suggested course of actions with already completed instances (cf. Fig. 6.10). The solid red line shows the execution of running process instance i_{r1} until the current execution state and the green × marks the current point in time. The dashed blue line, in turn, shows the predicted future course of actions of this instance. Table 6.1 depicts predicted values for test instances #3 and #5. The significant higher throughput time for test instance #5 results from repeating the actual task for handling the service request several times, whereas test instance #3 resolves the service request in the first run. Note that we consider other dimensions of process instances as well (i.e., throughput time and effort), i.e., we not only focus on the control flow.

Fig. 6.10 Predicted course: good versus bad

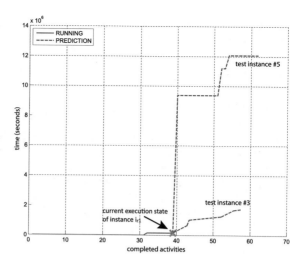

Table 6.1 Predicted values for instances i_{r1}

Instance	Throughput time	Effort
Test instance #3	21.74 days	20
Test instance #5	139.18 days	20

6.6   Related Work

A reference BPI architecture with several levels is presented in [16]. Level 1 contains *Data Integration* functions (e.g., connectors to different source systems), whereas Level 2 implements BPI core functions (e.g., KPI calculations). Level 3 comprises visualization components. Compared to this architecture, this chapter presented additional components, i.e., the *Prediction* and *Recommendation Engine* (both located on Level 2). In addition, this chapter emphasized the need of a *Process Control Engine*, which is required to propagate recommendations on the further course back to the PAIS.

Recommender systems are widely used in e-commerce platforms (e.g., Amazon), which suggest products to customers other customers previously purchased in a similar context [17, 18]. Usually, recommender systems are based on the review, recommendation and behaviour of other users [19]. In this context, [20] discusses issues related to the credibility of user-based recommender systems.

For analyzing log data, process mining algorithms exist [21]. For example, the *Decision Miner* algorithm calculates probabilities at decision points (i.e., XOR-split nodes) taking historical process executions into account [22]. For this purpose, decision trees are created and assigned to data within the process. Another approach is the *Recommendation Service* [23], which is provided for the declarative workflow system *Declare* [24]. Again, running process instances are checked against a set of already completed process instances. However, this approach does not take data element values, such as the effort to resolve a respective request, into account.

The approaches presented in [8, 25, 26] divide *predictive analysis* into *instance-based* (i.e., the prediction is made for a certain instance) and *class-based* (i.e., the prediction is made on basis of the corresponding class type) predictions. The approach presented in this chapter can be classified as a class-based prediction, since it compares a particular running process instance with already completed process instances (i.e., a cluster of instances). The approach described in [26] focuses on the time perspective (e.g., Can the given deadline be met?). Compared to this, the presented approach aims at integrating all available data of the given process instance (which are then used as different dimensions for our Edit Distance-Algorithm).

Finally, [27] uses decision trees to predict the KPIs of business processes. Based on KPI values of historic process instances, a prediction of the current instance is made.

6.7   Summary and Outlook

This chapter introduced an approach towards *predictive BPI*, which enables decision makers to predict the future progress of a business process instance enacted by a PAIS. Section 6.3 presented a methodology for predicting the course of business process instances as well as a corresponding system architecture. The focus was put on the *Prediction* and *Recommendation Engines* (cf. Sect. 6.4). We showed how established algorithms can be applied to discover similarities between process instances in order to predict the further course of execution of running process instances. Furthermore, we emphasized that not only information on the control flow of the process instance (e.g., executed activities) is analyzed in this context, but corresponding data element values (e.g., effort to handle the service request) need to be taken into account as well. This allows for a more accurate prediction of the ongoing execution of a process instance. In this context, an adaption of the Levenshtein algorithm for calculating distances between sequences of symbols was applied to derive first recommendations from this prediction.

Future research will integrate the prototype directly into a PAIS. Moreover, further algorithms will be added to realize more precise predictions and recommendations.

References

1. Reichert, M., Dadam, P.: Enabling adaptive process-aware information systems with ADEPT2. In: Handbook of Research on Business Process Modeling, pp. 173–203. Information Science Reference, Hershey, New York March (2009)
2. Reichert, M., Weber, B.: Enabling Flexibility in Process-Aware Information Systems: Challenges, Methods. Technologies, Springer, Berlin (2012)

3. Anandarajan, M., Anandarajan, A., Srinivasan, C.: Business Intelligence Techniques: A Perspective from Accounting and Finance. Springer (2004)
4. Doran, G.T.: There's a S.M.A.R.T. way to write management's goals and objectives. Manag. Rev. **70**(11), 35–36 (1981)
5. Eckerson, W.: Performance Dashboads: Measuring, Monitoring and Managing your Business, 2nd edn. Wiley Publishing Inc., Hoboken, New Jersey (2011)
6. Melchert, F., Winter, R., Klesse, M.: Aligning process automation and business intelligence to support corporate performance management. In: 10th Americas Conference on Information Systems, pp. 4053–4063 (2004)
7. Mutschler, B., Reichert, M., Bumiller, J.: Unleashing the effectiveness of process-oriented information systems: problem analysis, critical success factors and implications. IEEE Trans. Syst. **38**(3), 280–291 (2008)
8. Grigori, D., Casati, F., Castellanos, M., Dayal, U., Sayal, M., Shan, M.C.: Business process intelligence. Comput. Ind. **53**(3), 321–343 (2004)
9. van der Aalst, W.M.P., Weijters, A.J.M.M.: Process mining: a research agenda. Comput. Ind. **53**(3), 231–244 (2004)
10. Gangadharan, G.R., Swami, S.N.: business intelligence systems: design and implementation strategies. In: 26th International Conference on Information Technology Interfaces, 2004, pp. 139–144. IEEE (2004)
11. Fayyad, U., Uthurusamy, R.: Data mining and knowledge discovery in databases. Commun. ACM **39**(11), 24–26 (1996)
12. Lloyd, S.P.: Least squares quantization in PCM. IEEE Transcript Inf. Theor. **28**, 129–137 (1982)
13. Levenshtein, V.I.: Binary Codes Capable of Correcting Deletions, Insertions, and Reversals. Technical Report 8, Doklady Akademii Nauk (1966)
14. Dadam, P., Reichert, M.: The ADEPT project: a decade of research and development for robust and flexible process support—challenges and achievements. Comput. Sci. Res. Dev. **23**(2), 81–97 (2009)
15. MathWorks: MathWorks—MATLAB and Simulink. https://www.mathworks.com/ (2015). Accessed 20 Sep 2015
16. Mutschler, B., Bumiller, J., Reichert, M.: An approach to quantify the costs of business process intelligence. In: Proceedings of the International Workshop EMISA'05, pp. 152–163. No. P-75. Koellen-Verlag October (2005)
17. Linden, G., Smith, B., York, J.: Amazon.com recommendations: item-to-item collaborative filtering. IEEE Internet Comput. **7**(1), 76–80 (2003)
18. Schafer, J.B., Konstan, J., Riedi, J.: Recommender systems in e-commerce. In: Proceedings of the 1st ACM Conference on Electronic Commerce, pp. 158–166. EC '99, ACM, New York, NY, USA (1999)
19. Agrawal, R., Imieliński, T., Swami, A.: Mining association rules between sets of items in large databases. SIGMOD Rec. **22**(2), 207–216 (1993)
20. Cosley, D., Lam, S.K., Albert, I., Konstan, J.A., Riedl, J.: Is seeing believing?: how recommender system interfaces affect users' opinions. In: Proceedings of the SIGCHI Conference on Human Factors in Computing Systems, pp. 585–592. CHI '03, ACM, New York, USA (2003)
21. van Dongen, B., de Medeiros, A., Verbeek, H.M.W., Weijters, A.J.M.P., van der Aalst, W.M.P.: The ProM framework: a new era in process mining tool support. In: Applications and Theory of Petri Nets 2005, vol. 3536, pp. 1105–1116. Springer (2005)
22. Rozinat, A., van der Aalst, W.M.P.: Decision mining in ProM. In: Business Process Management, pp. 420–425. Springer (2006)
23. Schonenberg, H., Weber, B., Dongen, B., van der Aalst, W.M.P.: Supporting flexible processes through recommendations based on history. In: Proceedings of the 6th International Conference on Business Process Management, pp. 51–66. BPM '08, Springer (2008)
24. Pesic, M., van der Aalst, W.M.P.: A declarative approach for flexible business processes management. In: Business Process Management Workshops, pp. 169–180. Springer (2006)
25. Castellanos, M., Alves de Medeiros, A.K., Mendling, J.: Handbook of Research on Business Process Modeling, chap. Business Process Intelligence, pp. 456–480. IGI Global (2009)

26. Castellanos, M., Salazar, N., Casati, F.: Predictive business operations management. Int. J. Comput. Sci. Eng. **2**, 292–301 (2006)
27. Wetzstein, B., Zengin, A., Kazhamiakin, R., Marconi, A., Pistore, M., Karastoyanova, D., Leymann, F.: Preventing KPI violations in business processes based on decision tree learning and proactive runtime adaptation. J. Syst. Integr. **3**(1), 3–18 (2012)

Chapter 7
Reasoning About Process Models: What Description Logic Offers to Business Process Model Analysis

Michael Fellmann

Abstract Business process models are important for the design and implementation of process-aware information systems. Up to now, process models are represented predominantly as semi-formal models. Such models rely on the natural language to describe the models content via labels associated to the model elements. Due to the ambiguities of natural language, the semantics thus is not clear and well-defined. This in turn leads to problems when analyzing process models such as misinterpretations by humans or incomplete answers to queries by machines. In order to tackle this challenge, description logic-based ontologies provide well-defined semantics and can be used to represent graph-like knowledge structures such as business process models. Yet, up to now the capabilities of modern ontology languages are not widely used to represent, query and reason about process models. Therefore, the chapter presents an amalgamation of process models with ontologies. This amalgamation is formed by process models being represented in an ontology and being annotated with further elements of that ontology. In this way, the process model elements are augmented by machine processable semantics. By means of a concrete example, it is illustrated which deductions can be inferred using standard reasoning engines. With this, "intelligent" answers to queries executed against the knowledge base containing the process knowledge are possible that advance the model-based design of process aware information systems. Finally, an existing tool is briefly presented as a proof-of-concept. It allows creating and querying the ontology-based representation. This chapter is an excerpt of the introductory part of [1] which has been extended and revised.

M. Fellmann (✉)
Information Systems Research Group, University of Rostock,
Faculty of Computer Science and Electrical Engineering,
Albert-Einstein-Str. 22, 18059 Rostock, Germany
e-mail: Michael.Fellmann@Uni-Rostock.de

© Springer International Publishing AG 2017
G. Grambow et al. (eds.), *Advances in Intelligent Process-Aware Information Systems*, Intelligent Systems Reference Library 123, DOI 10.1007/978-3-319-52181-7_7

171

7.1   Introduction

For planning, controlling and managing of business processes and in order to handle the complexity associated with them, semi-formal models have been established. Typically, semi-formal modelling languages are used for the construction of such models. These languages try to balance mathematical accuracy with intuitive comprehension. Examples are the Business Process Model and Notation (BPMN), the Event-driven Process Chain (EPC) or the UML Activity Diagram. A characteristic feature of these languages is that the labels of model elements, e.g. *Check order* as a label of a BPMN task or EPC function, are assigned by the modeller with the help of natural language. Therefore, an essential part of the semantics of a process model is always bound to natural language. Due to the ambiguities of the natural language and a lack of (formalized) domain- or background knowledge, the processing of the models semantics is a challenging task.

The chapter at hand addresses this task. The semantics-related challenges of semi-formal modelling will be examined in more depth in Sect. 7.2. The process representation based on description logic which is the prerequisite for semantically well-defined model elements and machine reasoning is described in Sect. 7.3. How the knowledge contained in a description logic representation can be accessed via a query language is presented in Sect. 7.4. The types of inferred facts are characterized in Sect. 7.5. In order to use the concepts introduced, tool support is presented in Sect. 7.6. Finally, a conclusion is given in Sect. 7.7. This chapter is a revised and translated version of the introductory paper of [1] where the approach is described in more detail.

7.2   Semantics-Related Challenges of Semi-formal Modelling

In the following, some semantics-related challenges in semi-formal modelling are described. The first two challenges relate to the lack of unambiguous and machine processable semantics of individual model elements. The third challenge described is the lack of tools to support the creation and analysis of models with machine processable semantics at the level of individual model elements.

7.2.1   Ambiguities of the Natural Language

Natural language inevitably entails room for interpretation, which is, in the context of semi-formal modelling, referred to as a *linguistic defect* in literature. In this context, it is possible to distinguish between *synonyms, homonyms, equipollence, vagueness* and *incorrect designations* [2]. Especially models which are collaboratively created are problematic since agreeing on common terms is difficult in practice [3, 4].

The mentioned linguistic defects reduce the benefit of models as a medium of communication and emerged as one of the biggest problems of semi-formal modelling in practice [5]. For example, a reduced benefit or additional costs can emerge from synonyms in the labels of model elements. As a result, multiple drafts or multiple implementations of supporting information systems can occur in the subsequent phases [6]. Conversely, more recent research shows that commonly accepted and comprehended terminology are a factor of success for the development and implementation of information systems [7]. Furthermore, they cut costs, improve collaboration and simplify the decision-making for managers [8]. In this respect, projects are more successful when the actors early agree on a common terminology in comparison to projects where this is not the case [9].

7.2.2   Lack of Machine Processable Semantics

An inconsistent language used in conjunction with linguistic defects, as they occur in the current status quo of semi-formal modelling approaches, do not only complicate the interpretation and processing of models by human beings. They also prevent the (exact) processing of knowledge represented by the models with machines. However, this processing is essential in order to enable advanced process modelling tool features such as construction recommendations for model completion. Also, it is indispensable for automated process analyses on the level of the semantic model content concerning their completeness and a consistent degree of abstraction. Particularly for inexperienced modellers, modelling with a consistent degree of abstraction is challenging [10, 11]. Further, a problem while searching for models or model fragments is that the detection of facts and relations implicitly contained in the models is impossible although this information is deducible by logical conclusions. One example of this is a business process which includes a function that accesses resources stockpiled in a warehouse. Hence, the process reduces the stock. This deduction cannot be derived, if these connections are not specified in a machine processable form.

7.2.3   Lack of Semantics-Based Tool Support

Despite the variety of tools for generation, analysis and administration of models which were developed in the past, most of these tools do not consider the semantic content of individual model elements. Current advancements—especially in the commercial sector—mostly improve collaboration and cooperation aspects, but not the semantic support offered by the tool. This represents a gap in the current state of science and practice in particular against the background of the already developed standardized semantics in the form of extensive ontologies such as the MIT Process Handbook or the PCF taxonomy (Process Classification Framework). These are so far rarely used in tools for supporting model construction. Admittedly, in the area of Semantic Web Services there has been extensive research work in a scientific context to improve the transformation of (workflow) models to machine processable models [12–20]. But this research did barely lead to improved tools regarding the modelling of business-oriented process models.

Therefore, in the next Sect. 7.3 the fundamentals of an approach are introduced that enables to build advanced tools improving the machine support for construction and analysis of process models. This approach consists of a description logic-based process representation.

7.3   Description Logic-Based Process Representation

For representation purposes of formal ontologies, many languages in the area of artificial intelligence and especially in the area of the Semantic Web have been developed. The underlying description logics have been intensively researched for approximately 30 years. Semantic networks and frames can be thought of as precursors of description logic [21]. They intended a "natural" knowledge representation, while the efficiency of algorithms did not have priority. Contemporary description logics are designed with the aim to maintain an efficient computation despite a high expressiveness. Therefore, machine inference is also made possible within large knowledge bases (for the evolution of description logics, cf. [21, 22]).

In the past years, there has been a huge progress concerning the expressiveness and especially the scalability of knowledge bases. In this context, in particular the results of the "Billion Triple Challenge" are relevant—a contest where different developers and providers of knowledge base storages make a contest on processing data sets consisting of one billion triples (a triple is an elementary statement consisting of a subject, predicate and object—these terms a borrowed from linguistics). This mentioned progress has enabled the extension of semi-formal process modelling, which is described in this chapter.

The Web Ontology Language (OWL) is used for representation purposes within this research, because it is widespread even beyond the AI-research community. Further, it is standardized through the W3C [23] and a huge tool support is available. Specifically description logics based on the OWL-DL profile ("DL" therefore stands for "Description Logics") have been selected, because of the high expressiveness while retaining computational efficiency. There are powerful interference machines available for OWL-DL like *Pellet, FACT++* and *Hermit.*

In the following subsection, an overview of the approach is given from a conceptual view. Following this, an example is presented with an emphasis on representing the control flow of process models.

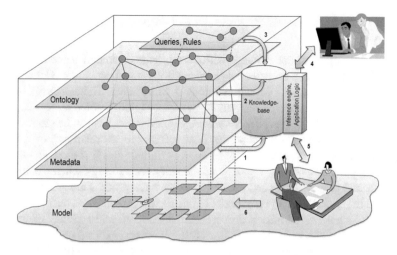

Fig. 7.1 Approach for semantic business process modeling

7.3.1 Conceptual Overview

The semantic process modelling presented in the context of this chapter is based on an ontology-based process model representation. The meta-model of the approach is described in [24]. Figure 7.1 illustrates essential elements of the approach for semantic business process modelling and their interaction.

The model-layer is connected with the layer of metadata by representing the model on the layer of metadata (dashed line between the layers). Thereby, the resulting generated metadata describing the model are stored in the knowledge base (arrow 1). They enable an interpretation of the model on the layer of ontology (arrow 2). This interpretation is possible because of the connections of elements from the metadata-layer with elements of the ontology-layer (lines between the metadata- and ontology-layer). In the context of this chapter, this connection is also named *semantic annotation*. On the layer of queries and rules, the possible inferences on the ontology-layer can be used to answer queries and check correctness conditions (arrow 3). They can not only relate to explicit represented, but also to logically deductible facts. The query and rules can both be used by analysts and model constructors (arrow 4 and 5) to retrieve information from the knowledge base and check the correctness. To do so, a user interface for example in the form of a modelling tool extension is required (cf. upper right image). The hereby possible insights can lead to a need of revising the model (arrow 6).

7.3.2   Ontology-Based Process Model Representation

In order to annotate model elements with a well-defined semantics and to query the process knowledge on a semantic level, an ontology-based process model representation is required. It will be explained briefly in this section. To begin with, in order to represent process models in an ontology, classes (also denoted as "concepts" in the context of ontologies) such as *Function*, *Event* or *Gate* have to be present in the ontology scheme. Moreover, properties (also referred to as "relations" or "connections") that connect instances (also referred to as "individuals") of those classes have to be specified such as *connects_to*. Fundamentally, the ontology schema in the form of classes and properties for representation of business process models was already presented in [25]. Thus in the remainder of this section, emphasis is put on the extension of this scheme by additional properties. In order to ease behavioural queries, the ontology schema was extended by a few OWL object properties (hereinafter simplified referred to as *properties*), which are used to represent the control flow in terms of behavioural queries. The following listing demonstrates the hierarchical structure of all properties of the extended ontology scheme by indentation.

```
graph_arc
   flow
     connects_to
       has_after_AND
       has_after_decision
          has_after_OR
          has_after_XOR
       has_after_event
       has_after_function
     flow_all
          flow_all_strict
     flow_strict
          flow_all_strict
     precedes
       precedes_all
   is_parallel_to
   is_exclusive_to
     is_exclusive_to_strict
   is_multichoice_to
     is_multichoice_to_strict
```

In the ontology language OWL, the property hierarchy is specified using the construct *rdfs:subPropertyOf* inherited from the RDF-schema. The addition of name space prefixes has been omitted in favour of a better readability. All properties are derived from the relation *graph_arc*. The transitive property *flow* specifies a directed path between model elements. A direct connection between two model elements is specified by a property *connects_to*. The type of the element following in the control flow can be implicitly indicated with semantically more specific properties. These comprise *has_after_AND*, *has_after_OR*, *has_after_XOR*, *has_after_ event* and *has_after_function*. As the property *flow* (such as all properties in RDF or OWL) is directed, a successor relationship is specified between two elements *a* and *b*, which are connected by *flow*. Predecessor relationships are not separately represented. If these are required for the specification of graph patterns in queries, they can be specified by interchanging the elements *a* and *b* in the query. Moreover, the terms of these properties are chosen in such a way so that they closely correspond with propositional logic operators in order to enable intuitive queries. These propositional logic operators are also eponymous for control flow operators in the EPC or other languages such as BPMN.

Moreover, the property *is_parallel_to* indicates a possible parallel execution. The properties *is_exclusive_to*, *is_exclusive_to_strict* and is *is_multichoice_to* as well as *is_multichoice_to_strict* indicate an exclusiveness relation.

In general, the suffix *strict* implies that the corresponding path is located outside of loops. The suffix *_all* occurring in properties such *flow_all* and *flow_all_strict* indicates that no alternative decisions are located on the path between two nodes. Paths with the properties *precedes* and *precedes_all* are interrupted when branches caused by alternative decisions are merged. This is done because from the perspective of the corresponding join connector, it cannot be specified which elements were previously executed. The property *precedes_all* is also interrupted before loops, because elements can be activated multiple times in loops without the element before the loop to be executed.

Figure 7.2 illustrates the existence of the properties by a process example. In this example, the node $F10$ is selected. All other nodes in the model which are connected to $F10$ via chains of properties are marked with symbols representing the respective type of the properties. For example, all tokens that pass node $F10$ also pass $F9$, $E3$ and $E2$—however, it may be the case that $F9$ is executed before $F10$ or vice versa. A node that is executed strictly after $F10$ and executed for all tokens that pass $F10$ is $E2$.

Hierarchical structuring of the properties for control flow representation enables queries with different accuracy. This is also requested from Beeri et al. [26] for an adequate query language. For instance, with the property *flow* it can be discovered whether a connection between two elements exist. Further, with the property *has_after_XOR* only such model elements in the ontology-based representation can be found which directly follow a logical alternative decision. The featured properties in queries can be combined arbitrarily. In this way, some parts of queries may be more precisely while others are coarser enabling a flexible querying of process knowledge.

The presented ontology-based representation of process models can be generated using the algorithm described in [1] for so called "structured models" only. That is, the models must contain balanced splitting and merging connectors whereby each opening connector is closed with a connector of the same type so that blocks of nested

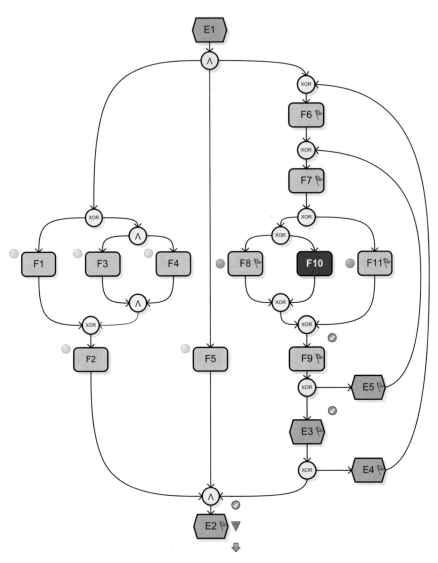

Legend:

⚐ = flow, ✓ = flow_all, ▼ = flow_all_strict, ⬇ = flow_strict, ◯ = parallel_to, ● = exclusive_to

Fig. 7.2 Sample model with ontology properties for control flow representation

connectors occur. However, the models are allowed to contain do-while-loops (such as in Fig. 7.2 between $F6$ and $E3$). If the models are unstructured, a subset of the properties can be generated. To do so, the behavioural profiles described and implemented by [27] can be used. Using these profiles, three relations can be computed that are strict order, exclusiveness and interleaving order. The first two can be used to detect the `flow_strict` and `is_exclusive_to_strict` properties.

7.4 Querying Process Knowledge

In this Section it is explained how the knowledge captured in the ontology-based process model representation (cf. Sect. 7.3) can be queried using the standard semantic web query language SPARQL. This language is a standardized query language by the W3C to query graph structures. It achieved a de facto status especially in the area of Semantic Web and the Linked Open Data (LOD)-movement [28]. This success may be caused by the simple basic structure of the queries. SPARQL is meanwhile supported by major commercial databases of the business environment. These also offer a scalable storage of large quantities of OWL ontologies or RDF files as well as manifold inference capabilities. The relevance of this aspect is emphasized by real world enterprises, which use hundreds of models [29], which achieve a substantial size (the so-called "process wallpaper") [30].

A SPARQL query refers to RDF triples, which represent a directed graph. Basically, the query consists of a *pattern matching*, a *modifier of solutions* and a *return* part [28]. The function of *pattern matching* is to compare a graph pattern, which is composed out of triple patterns, against an RDF-graph. In the context of the ontology-based process representation, graph patterns can be intuitively understood as structure or "a way through the graph". The triple patterns, which determine the graph pattern, represent the navigation step. At the position of the subject and object of a pattern, classes of ontology or the instances of the ontology can be specified. At the position of the predicate, the ontology properties can be specified. Moreover, at the position of the object, simple data values such as literals (e.g. character strings) can be specified. A simple query to return all instances in the ontology, which are connected through a property `predicate`, can be formulated as follows.

`SELECT ?var1 ?var2 WHERE { ?var1 predicate ?var2 }`

Within this chapter, this general scheme will be applied to query process models. Due to the chosen representation of the process models, in which the elements of the model are represented as instances and the control flow as properties in the ontology, an intuitive application of SPARQL is possible for searching in process graphs. The represented elements of the model in the ontology occur in these queries at the position of the subject or the object in a triple pattern. By combining several patterns, complex queries can be created. Queries may also contain placeholders, so called blank- or *anonymous nodes*. A simple example for this would be to return all activities being followed by a XOR decision.

`SELECT ?decision WHERE { ?decision :has_after_XOR [] }`

The graph model of this query defines one single triple pattern; its subject is the variable `?decision`, which is also used for the return, its predicate has the property `:has_after_XOR` and its object is an anonymous node [] specified as a dummy for arbitrary nodes. Due to the integration of behavioral properties like

:flow_all_strict in the ontological representations, behavioral queries are possible as well. For example all activities, which in any case will be executed once after the process is started, can be retrieved by the following query.

```
SELECT ?mandatory WHERE  {
    ?x a :StartEvent ; :flow_all_strict ?mandatory
}
```

For queries with complex paths between two nodes, *property path expressions* enable a significant gain in expressiveness and likewise a simplified notation. For example to return such model element pairs, which are on a path connected with at least one subsequent XOR- or OR-split and after this with at least two sequenced AND-splits, the following query can be used.

```
SELECT ?node1 ?node2 WHERE {
    ?node1 ( :has_after_XOR | :has_after_OR )+
             / :has_after_AND{2,}/:connects_to ?node2
}
```

Path expressions allow the declaration of cardinalities, in short form for arbitrary *, for at least one + and optional ? or in detail by a min-max-notation {min,max}. The concatenation of several properties for the navigation through the graph can be written with a slash / and alternatives are represented in parentheses (opt1 | opt |...| optN). However, with SPARQL it is not possible to return the specific pattern, including the nodes, when not only the nodes at the end of a path are relevant. This can easily be upgraded with extensions like GLEEN [31]. Other extensions of SPARQL allow more comprehensive restrictions for searching on a graph. For example, these restrictions can refer to the presence or absence of specific nodes on a path. With RPL (RDF Path Language) [32] an approach exists for the navigation in RDF-graphs, similar to XPath expressions in the context of XML-standards. SPARQ2L supports the subgraph extractions from RDF-data and with CSPAROL (Constrained SPARQL) [33] an extension for describing path-restrictions exists, which can be explained in cooperation with the extension PSPARQL (Pattern SPARQL) [34]. Sometimes also new query languages like RDFPath are developed, which enable an expressive path-declaration on large RDF graphs [35]. More research that could be applied to extend the approach presented here exists in the area of graph databases [36].

To reflect the semantic annotation in the query, the graph model has to be extended with corresponding triple patterns, specifying the annotation. For example to return all model elements that are annotated with an ontology instance :notifiy_customer via the annotation property :equivalent_to, the following query can be used.

```
SELECT ?notify WHERE {
    ?notify :equivalent_to :notify_customer
}
```

A more abstract form of a query occurs when the ontology-instance used to annotate a model element is unknown. Therefore an ontology-class will be specified
instead of an ontology-instance. In addition, the annotation can be formulized more
unspecific, by using the super-property :has_annotation instead of the sub-
properties :equivalent_to and :narrower_than. In this way, annotations
that are semantically equivalent as well as more general will be found. For example to return all model elements that are annotated with an ontology-instance of the
class :StrategicActivity via the property :has_annotation, the following query can be used.

```
SELECT ?strategic WHERE {
    ?strategic :has_annotation [ a :StrategicActivity ]
}
```

By querying the ontology-instance type with a, which is a short form for
rdf:type, the entire spectrum of conclusions for classifying instances, that modern description logics like OWL 2 for the classifications of instances provide, can be
used. This machine inferred facts will be included in the result of the query together
with other inferences for example based on transitive properties.

In the next section, the deductions that are added to the ontology-based process
model representation by standard reasoners are described. The deductions result in
additional facts being inferred by machine reasoning procedures. These facts enrich
the results that can be retrieved by queries such as described in this chapter. In the
next chapter, the facts are characterized in further detail.

7.5 Use of Machine Reasoning

In this section, at first an overview on the subject of automatic reasoning is provided.
After this, a characterization of important types of inferences is provided. Inferences
are distinguished first according to whether they relate to type information or relation information. Second, according to whether they involve the ontology instances
representing the process model or the domain. These distinctions lead to a matrix
of four inference types (cf. Fig. 7.3 bottom). Examples to explain these inference
types are subsequently illustrated by (a) pointing to a coherent annotated sample
process graphically illustrated in Fig. 7.3, (b) by describing the inference type and
its practical value for querying using natural language and (c) by giving an example
in description logics syntax along with explanations.

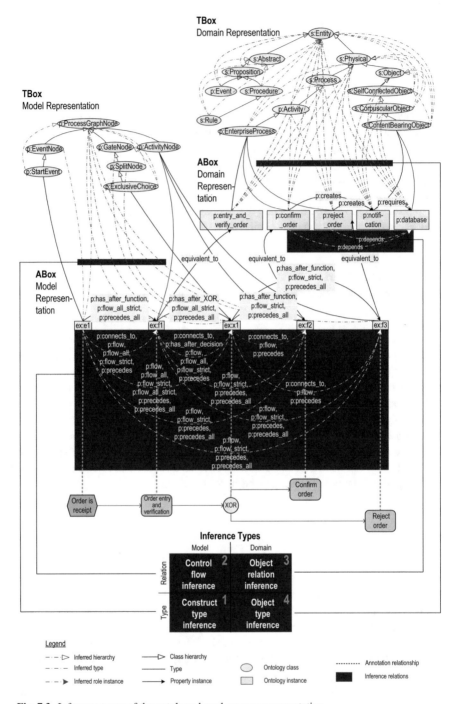

Fig. 7.3 Inference types of the ontology-based process representation

7.5.1 Overview

In this section, types of machine inferences will be characterized. The types are intended and described in terms of their relevance for the business-level interpretation of process models, not as general categories of inferences possible with description logics. Figure 7.3 illustrates the types that are connected to the respective inferences depicted by a line with a dot-and-dash-pattern integrated in the sample ontology-based process representation. In more detail, the classification of an inference to an inference type follows out of the crossing of such lines through a dark filled rectangle, which is connected to a corresponding inference type. The namespace `p:` used in Fig. 7.3 stands for the extensions of the SUMO-ontology [37], `s:` for the SUMO-ontology itself and `ex:` (example) for any example-namespace. The inferences presented in Fig. 7.3 represent additions both to the TBox (Terminological Box, also called *ontology scheme*) and the ABox (Assertional Box, also called *ontology data*) of the ontology. In this contribution, the term *ontology* includes both TBox and ABox. This use is common in the context of OWL ontologies.

7.5.2 Characterization of the Four Types of Inferences

According to their content, the conclusions can be classified into four inference types being *construct type inference*, *control flow inference*, *object relation inference* and *object type inference*. They will to be characterized in the following.

7.5.2.1 Construct Type Inference

Concerned here are conclusions that relate to the *type* of ontology instances which represent the model elements. In queries, the *construct type inference* enables an abstraction of the ontology classes which are used to represent the constructs of a modelling language. For example, the fact that the ontology instance `ex:x1` (the XOR-connector) is inferred to be of type `p:GateNode` is a construct type inference since the type of the model element is inferred. In Fig. 7.3, this inference is depicted by the dotted line between the ontology instance belonging to the ABox model representation and the corresponding ontology class belonging to the TBox model representation. In addition to the graphical illustration, this inference is explained in the following using the DL-syntax common in description logics. The descriptions are divided into two parts: An inferred fact and its explanation.

Inferred fact:
   $p: GateNode(ex:x1)$

Explanation:
   $p: ExclusiveChoice(ex:x1)$
   $p: ExclusiveChoice \sqsubseteq p : SplitNode$
   $p: SplitNode \sqsubseteq p: GateNode$

The inferred fact expresses that the individual `ex:x1` is a member of the class `p:GateNode`. This can be explained by the membership of `ex:x1` in the class `p:ExclusiveChoice` being a subclass of `p:SplitNode` which in turn is a subclass of `p:GateNode`.

An example for the practical use of the construct type inference in queries is the search for events, without the need to specify whether it is a start, intermediate or end event. Without this type of inference, the desired type of event has to be specified exactly or all sorts have to be enumerated in the query. Hence using construct type inference, a variable degree of abstraction from the constructs of the modelling language originally used for constructing the model can be achieved.

7.5.2.2   Control Flow Inference

These are conclusions relating to the properties in the ontology representing the control flow of the process model. In queries, the *control flow inference* provides an abstraction of control flow structures of a process model. For example, the fact that `ex:e1` (Order is receipt) is inferred to be connected via a `p:flow`-property with `ex:f2` (Confirm order) is a control flow inference since two elements that are not directly connected are inferred to be connected via properties. In Fig. 7.3, this inference is depicted by the arrow with a dotted line and a label "p:flow" between the two ontology instances in the ABox model representation. In DL-Syntax, the example is described as shown below.

Infered fact:
   $p: flow(ex:e1, ex:f2)$

Explanation:
   $p: flow\_strict(ex:e1, ex:f1)$
   $p: flow\_strict(ex:f1, ex:x1)$
   $p: flow\_strict(ex:x1, ex:f2)$
   $p: flow\_strict^+ \equiv p: flow\_strict$
   $p: flow\_strict(ex:e1, ex:f2)$
   $p: flow\_strict \sqsubseteq p: flow$

The inferred fact expresses that the individual ex:e1 has a property p:flow with a value ex:f2, in other words, that ex:e1 and ex:f2 are connected by the property p:flow. This can be explained by the three connections ex:e1 with ex:f1, ex:f1 with ex:x1 and ex:x1 with ex:f2 via the property p:flow_strict. Since the property is transitive (which is indicated using the symbol "+" in the DL-syntax), it follows that ex:e1 is also connected to ex:f2. Since the property p:flow_strict is a sub-property of p:flow, it follows that ex:e1 is connected to ex:f2 via p:flow.

The practical use of the control flow inference lies in the abstraction of concrete structures in queries, because the inferred properties in the ABox allow to query "direct connections" between the represented model elements in the ontology. Moreover, using the property hierarchy in the TBox of the ontology, a variable degree of abstraction of the control flow can be achieved.

7.5.2.3 Object Relation Inference

Conclusions of this type refer to the relations between ontology instances used for annotating model elements and the domain representing instances in the ontology. The *object relation inference* allows to request more context and information about annotated model elements that are not explicitly covered in the ontology, but can be inferred by the inference engine. In particular by transitive properties or property chains diverse conclusions may result which lead to an advanced semantic interpretation of an annotated model element and thus to new insights. For example, the fact that the instance p:confirm_order is dependent on the p:database is an object relation inference since a relation between an annotated model element and another object from the domain is inferred. The inferred fact is also depicted in Fig. 7.3 by the arrow with a dotted line and a label "p:depends" between the two instances in the ABox domain representation. In DL-Syntax, the example is described in more detail containing the relevant background knowledge as shown below.

Inferred fact:
$$p{:}depends(p{:}confirm\_order, p{:}database)$$

Explanation:
$$p{:}creates(p{:}confirm\_order, p{:}notification)$$
$$p{:}requires(p{:}notification, p{:}database)$$
$$p{:}createsorequires \rightarrow depends$$

The inferred fact expresses that p:confirm_order has a property p:depends with a value p:database, in other words, that p:confirm_order is connected to p:database by the property p:depends. This can be explained by two properties and a property chain. The first property p:creates connects

p:confirm_order with p:notification, the second property
p:requires connects p:notification with p:database. Since there is
a property chain specified (last line in the explanation-part of the DL-syntax above)
that says that the composition of the properties p:creates and p:requires
constitutes a p:depends-relation, it can be inferred that p:confirm_order
and p:database are connected by the property p:depends.

The practical use of object relation inferences in queries lies for example in discovering dependencies that are imposed on the execution of activities or the detection
of objectives (transitively) supported by an activity.

7.5.2.4   Object Type Inference

Conclusions of this type relate to the type of instances in the ontology which represent
domain knowledge and that are used to annotate model elements. In queries, the
object type inference enables the abstraction of these concrete ontology instances
used for annotation. For example, the fact that p:entry_and_verify_order
is of type s:Process is an object type inference since the type of the instance
p:entry_and_verify_order which is used to annotate the activity ex:f1 is
inferred. In Fig. 7.3, this inferred fact is also depicted by the dotted line between the
ontology instance in the ABox domain representation and the respective ontology
class in the TBox domain representation. In DL-Syntax, the example is described as
shown below.

Inferred fact:
   $s: Process(p: entry\_and\_verify\_order)$

Explanation:
   $p: Enterprise Process(p: entry\_and\_verify\_order)$
   $p: Enterprise Process \sqsubseteq p: Activity$
   $p: Activity \sqsubseteq s: Process$

The inferred fact expresses that the individual p:entry_and_verify_order
is a member of the class s:Process. This can be explained by the membership of
this individual in the class p:EnerpriseProcess being a subclass of the more
general class of activities p:Activity which in turn is a subclass of s:Process.

A practical example for the application of this type of conclusion in queries
would be the search for all nodes in the process graph that are annotated with an
ontology instance that creates a document type which is an instance of the class
ContentbearingObject. As a result of the query, the exact document type
such as fax, e-mail, or letter, which can be defined as subclasses, are abstracted by
means of the object type inference.

Finally, the inferences of the object relation inference and the object type inference are heavily dependent on the descriptions and richness of the ontology. The scope and content of the ontology should be determined according to the questions to be answered, which are also called *competence questions* in the area of knowledge-based systems [38]. Examples of competence questions would be "What organizational unit is responsible for an activity?", "Which employee performs which task?", "What best practices apply for a job?", "Which cost causes the execution of an activity?", "What guidelines need to be followed?", "Which national category corresponds to a process?", just to mention a few examples. The questions to be answered influence the design of the ontology.

7.6 Tool Support

In the following, we describe the tool that has been developed for querying the ontology-bases process model representation. The following description and screen-shots are based on [39] where the tool is described in more detail in [1]. The tool is called *SemQuu*—Semantic Query Utility, since much of the functionality is also relevant when exploring ontologies in general. SemQuu is implemented using the Jena library (jena.sourceforge.net), the Pellet inference engine (pellet.owldl.com), the Tomcat web server and JSP, XSLT, JavaScript, CSS and HTML for the user interface. Process models can be imported from arbitrary tools in arbitrary formats, since they can be transformed on the server into the ontology-based representation which is accomplished by a plugin-converter for each format. As an example, we have implemented an extension of Visio which can export EPC process models being annotated with ontology concepts (see Fig. 7.4 for an illustration of the following procedure). After having been exported from Visio to SemQuu, the model is trans-formed to an OWL-DL ontology and added to the repository (cf. Fig. 7.4 bottom right).

An overview of SemQuu is provided by Fig. 7.5 (B). In order to query the repos-itory of ontology-based process representations with SPARQL, the user can use a simple form-based query builder (A) for successively constructing a graph pattern. This is done by inserting multiple triple patterns with the help of drop-down list boxes (A) which are aware of `rdfs:domain` and `rdfs:range` information so that no semantically wrong query can be constructed. Moreover, drop-down list boxes are dynamically updated upon selection of a value in one of the boxes. Alternatively to the drop-down list boxes, the user can leverage the full expressivity of SPARQL by using the text area input field. Moreover, when the user modifies the query she or he is supported by an "intelligent" auto-completion feature (C) which is fully aware of the ontology schema and instance data and only proposes meaningful suggestions. When queries are executed in batch mode, the result of the queries can be displayed as an information, warning or error with respective graphical symbols appearing for each type (D). The result for each query is initially collapsed but can be unfolded if the user clicks on the "+" sign symbols. In order to measure the effectiveness of

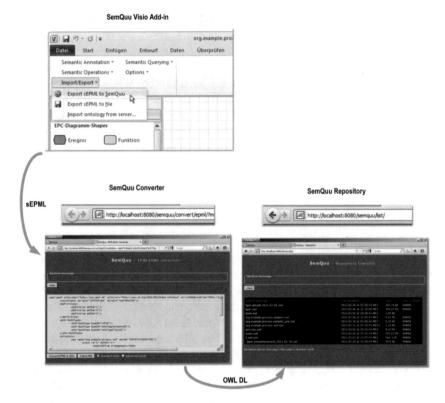

Fig. 7.4 SemQuu converter and repository

SPARQL queries for the task of semantic correctness checking, we conducted an experiment with 21 participants described in [40].

7.7 Discussion and Outlook

Using a description logics-based ontological process representation enables a well-defined semantics for model elements that at the same time is also machine processable. The querying against ontology-based process representations enables a comprehensive analysis of process models. Structure-related as well as behavioural queries can be answered through the control flow representation. Relationships between processes, as they are possible in BPMN, EPC or other languages e.g. via process interfaces or sub-processes are currently not taken into account yet, but in the future, they can be regarded by introducing additional properties such as starts_subprocess and returns_to_process in the ontology-based process modeling.

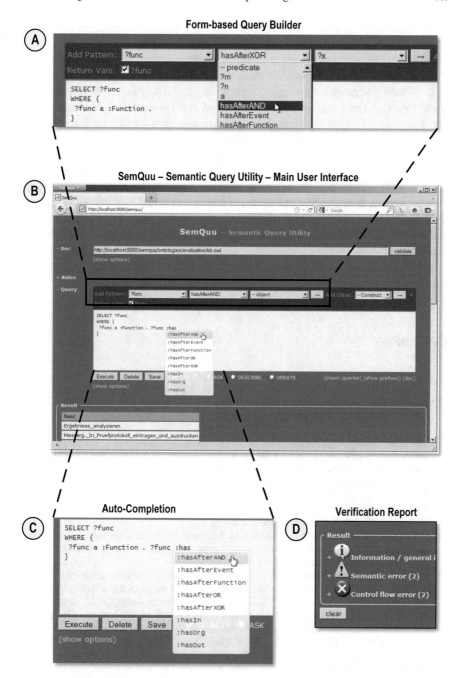

Fig. 7.5 Overview of SemQuu

In contrast to the existing approaches, the approach presented here allows the integration of the full spectrum of possible deductions with descriptions logics like OWL-DL in the results of a query, which enables the extensive use of "mechanically" created conclusions. This spectrum has been described by four classes of inference types. While the *construct type inference* and the *control flow inference* refer to inferences based on the structure of the represented process model, the *object relation* inference and the *object type inference* refer to the domain knowledge formalized in the ontology. With the presented approach of ontology-based process representation, it is possible to seamlessly combine these two knowledge areas and to reason about the process representation using description logics inference engines. Thus, a richer analysis and interpretation of the organizational process knowledge can be achieved. Future research in this field can examine a refinement of inference types or a quantification and measurement of its occurrence and usefulness in large process model repositories. Another direction of research is the automated annotation of process models (or at least, creating suggestions for annotation in an automated way). Finally, evolving the ontology collaboratively and at the same time keeping the annotation intact is subject to future research.

References

1. Fellmann, M.: Semantic Process Engineering – Konzeption und Realisierung eines Werkzeugs zur semantischen Prozessmodellierung, PhD thesis. Osnabrück University, Osnabrück (2013)
2. Ortner, E.: Methodenneutraler Fachentwurf: Zu den Grundlagen einer anwendungsorientierten Informatik. Teubner-Reihe Wirtschaftsinformatik. Teubner, Stuttgart (1997)
3. Scheer, A.W., Klueckmann, J.: BPM 3.0. In: Dayal, U., Eder, J., Koehler, J., Reijers, H.A. (eds.) Proceedings of the Business Process Management 7th International Conference (BPM 2009), 8–10 Sept, Ulm, Germany. LNCS, vol. 5701, pp. 15–27. Springer, Berlin (2009)
4. Hadar, I., Soffer, P.: Variations in conceptual modeling: classification and ontological analysis. J. Assoc. Inf. Syst. 7(8), 568–592 (2006)
5. Sarshar, K., Weber, M., Loos, P.: Einsatz der Informationsmodellierung bei der Einführung betrieblicher Standardsoftware: Eine empirische Untersuchung bei Energieversorgerunternehmen. Wirtschaftsinformatik 48(2), 120–127 (2006)
6. Rosemann, M.: Komplexitätsmanagement in Prozessmodellen: Methodenspezifische Gestaltungsempfehlungen für die Informationsmodellierung. Schriften zur EDV-orientierten Betriebswirtschaft, Gabler, Wiesbaden (1996)
7. Corvera, M., Rosenkranz, C.: Natural language alignment as a process: applying functional pragmatics in information systems development. In: Proceedings of the 18th European Conference on Information Systems (ECIS 2010), 6–9 June, Pretoria, South Africa. Paper 5 (2010)
8. Schafermeyer, M., Grgecic, D., Rosenkranz, C.: Factors influencing business process standardization: a multiple case study. In: Proceedings of 43rd Hawaii International Conference on System Sciences (HICSS 2010), 5–8 Jan, Poipu, Kauai, Hawaii, pp. 1–10. IEEE (2010)
9. Rosenkranz, C., Räkers, M., Behrmann, W., Holten, R.: Supporting financial data warehouse development: a communication theory-based approach. In: Proceedings of the Thirty First International Conference on Information Systems (ICIS 2010), 12–15 Dec, Saint Louis, Missouri, USA. Paper 12 (2011)
10. Nielen, A., Költer, D., Mütze-Niewöhner, S., Karla, J., Schlick, C.: An empirical analysis of human performance and error in process model development. In: Proceedings of the 30th

International Conference on Conceptual Modeling (ER 2011), Brussels, Belgium, pp. 514–523 (2011)

11. Wilmont, I., Brinkkemper, S., Weerd, I., Hoppenbrouwers, S.: Exploring intuitive modelling behaviour. In: Bider, I., Halpin, T., Krogstie, J., Nurcan, S., Proper, E., Schmidt, R., Ukor, R. (eds.) Enterprise, Business-Process and Information Systems Modeling: Proceedings of the 11th International Workshop, BPMDS 2010 and 15th International Conference, EMM-SAD 2010 held at CAiSE 2010, 7–8 June, Hammamet, Tunisia. LNBIP vol. 50, pp. 301–313. Springer, Berlin (2010)

12. Cabral, L., Domingue, J., Motta, E., Payne, T.R., Hakimpour, F.: Approaches to semantic web services: an overview and comparisons. In: Bussler, C., Davies, J., Fensel, D., Studer, R. (eds.) The Semantic Web: Research and Applications: Proceedings of the First European Semantic Web Symposium, ESWS 2004 Heraklion, Crete, Greece, 10–12 May. LNCS, vol. 3053, pp. 225–239. Springer, Berlin (2004)

13. Cardoso, J., Sheth, A.P.: Introduction to semantic web services and web process composition. In: Cardoso, J., Sheth, A.P. (eds.) Semantic Web Services and Web Process Composition: First International Workshop, SWSWPC 2004, San Diego, CA, USA, 6 July, pp. 1–13 (2005)

14. Roman, D., Keller, U., Lausen, H., de Bruijn, J., Lara, R., Stollberg, M., Polleres, A., Feier, C., Bussler, C., Fensel, D.: Web service modeling ontology. Appl. Ontol. 1(1), 77–106 (2005)

15. The OWL Services Coalition (eds.) OWL-S: Semantic Markup for Web Services

16. Farell, J., Lausen, H. (eds.): Semantic Annotations for WSDL: W3C Recommendation, 28 Aug 2007. W3C (2007)

17. Wetzstein, B., Ma, Z., Filipowska, A., Kaczmarek, M., Bhiri, S., Losada, S., Lopez-Cob, J.-M., Cicurel, L.: Semantic business process management: a lifecycle based requirements analysis. In: Hepp, M., Hinkelmann, K., Karagiannis, D., Klein, R., Stojanovic, N. (eds.) Proceedings of Workshop on Semantic Business Process and Product Lifecycle (SBPM 2007) in conjunction with the 4th European Semantic Web Conference (ESWC 2007), 3–7 June, Innsbruck, Austria. CEUR Workshop Proceedings, vol. 251, pp. 1–11. RWTH, Aachen (2007)

18. Weber, I.M.: Verification of annotated process models. In: Weber, M. (ed.) Semantic Methods for Execution-level Business Process Modeling, Part 2. LNBIP, vol. 40, pp. 97–148. Springer, Berlin (2009)

19. Weber, I., Hoffmann, J., Mendling, J.: Beyond soundness: on the verification of semantic business process models. Distrib. Parallel Databases 2010(27), 271–343 (2010)

20. Drumm, C., Filipowska, A., Hoffmann, J., Kaczmarek, M., Kaczmarek, T., Kowalkiewicz, M., Markovic, I., Scicluna, J., Vanhatalo, J., Völzer, H., Weber, I., Wieloch, K., Zyskowski, D.: Dynamic Composition Reasoning Framework and Prototype. Project IST 026850 SUPER, Deliverable 3.2. SAP (2007)

21. Baader, F.: What's new in description logics. Informatik-Spektrum 34(5), 1–9 (2011)

22. McGuinness, D.L.: Description logics emerge from ivory towers. In: Goble, C.A., McGuiness, D.L., Möller, R., Patel-Schneider, P.F. (eds.) Proceedings of the International Workshop on Description Logics, 1–3 Aug, Stanford University, California, USA, pp. 64–68 (2001)

23. OWL Working Group: OWL 2: W3C Recommendation, 11 Dec 2012 (2012)

24. Thomas, O., Fellmann, M.: Semantische Prozessmodellierung – Konzeption und information-stechnische Unterstützung einer ontologiebasierten Repräsentation von Geschäftsprozessen. Wirtschaftsinformatik 51(6), 506–518 (2009)

25. Thomas, O., Fellmann, M.: Semantic process modeling—design and implementation of an ontology-based representation of business processes. Bus. Inf. Syst. Eng. 1(6), 438–451 (2009)

26. Beeri, C., Eyal, A., Kamenkovich, S., Milo, T.: Querying business processes. In: Dayal, U., Whang, K.Y., Lomet, D.B., Alonso, G., Lohman, G.M., Kersten, M.L., Cha, S.K., Kim, Y.K. (eds.) Proceedings of the 32nd International Conference on Very Large Data Bases (VLDB 2006), 12–15 Sept, Seoul, Korea, pp. 354–366. VLDB Endowment, ACM (2006)

27. Weidlich, M., Polyvyanyy, A., Mendling, J., Weske, M.: Causal behavioural profiles—efficient computation, applications, and evaluation. Fundamenta Informaticae (FI) 113(3–4), 399–435 (2011)

28. Pérez, J., Arenas, M., Gutierrez, C.: Semantics and complexity of SPARQL. ACM Trans. Database Syst. (TODS) **34**(3), 16 (2009)

29. Awad, A.: BPMN-Q: A language to query business processes. In: Reichert, M., Strecker, S., Turowski, K. (eds.) Proceedings of the 2nd International Workshop on Enterprise Modelling and Information Systems Architectures (EMISA2007), 8–9 Oct, St. Goar, Germany, pp. 115–128 (2007)

30. Polyvyanyy, A., Smirnov, S., Weske, M.: Business process model abstraction. In: vom Brocke, J., Rosemann, M. (eds.) Handbook on Business Process Management 1: Introduction, Methods, and Information Systems: Part II Methods. International Handbooks on Information Systems, pp. 149–166. Springer, Berlin (2010)

31. Detwiler, L.T., Suciu, D., Brinkley, J.F.: Regular paths in SparQL: querying the NCI thesaurus. In: Proceedings of the American Medical Informatics Association Symposium on Biomedical and Health Informatics (AIMA 2008), 8–12 Nov, Washington, DC, pp. 161–165 (2008)

32. Zauner, H., Linse, B., Furche, T., Bry, F.: A RPL through RDF: expressive navigation in RDF graphs. In: Hitzler, P., Lukasiewicz, T. (eds.) Proceedings of the 4th International Conference on Web Reasoning and Rule Systems (RR 2010), 22–24 Sept, Bressanone/Brixen, Italy. LNCS, vol. 6333, pp. 251–257. Springer, Berlin (2010)

33. Alkhateeb, F., Baget, J.F., Euzenat, J.: Constrained regular expressions in SPARQL: Proceedings of the International Conference on Semantic Web and Web Services (SWWS), 14–17 July, Las Vegas, NV US, pp. 91–99 (2008)

34. Alkhateeb, F., Baget, J.F., Euzenat, J.: Extending SPARQL with regular expression patterns (for querying RDF). Web Semant. Sci. Serv. Agents World Wide Web **7**(2), 57–73 (2009)

35. Przyjaciel-Zablocki, M., Schätzle, A., Hornung, T., Lausen, G.: RDFPath: path query processing on large RDF graphs with MapReduce. In: Proceedings of the 1st Workshop on High-Performance Computing for the Semantic Web (HPCSW2011) co-located with the 8th Extended Semantic Web Conference, ESWC2011, 29 May, Heraklion, Greece (2011)

36. Angles, R., Gutierrez, C.: Querying RDF data from a graph database perspective. In: Gómez-Pérez, A., Euzenat, J. (eds.) Proceedings of the Second European Semantic Web Conference, ESWC 2005, Heraklion, Crete, Greece, May 29–June 1. LNCS, vol. 3532, pp. 346–360. Springer, Berlin (2005)

37. Niles, I., Pease, A.: Towards a standard upper ontology. In: Welty, C., Smith, B. (eds.) Proceedings of the 2nd International Conference on Formal Ontology in Information Systems (FOIS-2001), 17–19 Oct, Ogunquit, Maine, pp. 2–9 (2001)

38. Gómez-Pérez, A., Fernández-López, M., Corcho, O.: Ontological Engineering: With Examples from the Areas of Knowledge Management, E-Commerce and the Semantic Web. Springer, London (2004)

39. Fellmann, M., Thomas, O.: Process model verification with SemQuu. In: Nüttgens, M., Thomas, O., Weber, B. (eds.) Enterprise Modelling and Information Systems Architectures (EMISA 2011), 22–23 Sept, Hamburg, Germany. GI LNI P-190, pp. 231–236. Köllen, Bonn (2011)

40. Fellmann, M., Thomas, O., Busch, B.: A query-driven approach for checking the semantic correctness of ontology-based process representations. In: Abramowicz, W. (ed.) Proceedings of the 14th International Conference on Business Information Systems (BIS 2011), 15–17 June, Poznan, Poland. LNBIP, vol. 87, pp. 62–73. Springer, Berlin (2011)

Chapter 8
Improving Process Portability Through Metrics and Continuous Inspection

Jörg Lenhard

Abstract Runtimes for process-aware applications, i.e., process engines, constantly evolve and in the age of cloud-enabled process execution, the need to change a runtime quickly becomes even more evident. To cope with this fast pace, it is desirable to build processes in a way that makes them easily portable among engines. Reliance on process standards is a step in the right direction, but cannot solely solve all problems. Standards are just specifications from which implementations will naturally deviate, thus fueling the problem of process portability. Here, the field of software measurement can provide some remedy. Metrics for process portability can help to make intelligent decisions on whether to invest in porting or rewriting process-aware applications. What is more, if integrated into the development process through agile techniques like continuous inspection, portability metrics can help in the implementation of more portable processes from the very beginning. In this chapter, we present an approach for the measurement of process portability and explain how this can improve decision making and process quality in general. The approach builds on the recently revised version of the renowned ISO/IEC software quality model and we describe how this model is in line with techniques of continuous inspection. We discuss what constitutes process portability and present a set of newly proposed software metrics for quantifying portability.

Keywords Portability · Process quality · Metrics · Continuous inspection

8.1 Why Process Portability Matters

It has never been easier to provision a new and scalable runtime environment for an application than today. In the times of cloud computing, new computing resources can be acquired on demand and set up within seconds. This enables applications to scale up and down intelligently, depending on the workload put onto the system [1].

J. Lenhard (✉)
Department of Mathematics and Computer Science, Karlstad University, Karlstad, Sweden
e-mail: joerg.lenhard@kau.se

© Springer International Publishing AG 2017
G. Grambow et al. (eds.), *Advances in Intelligent Process-Aware
Information Systems*, Intelligent Systems Reference Library 123,
DOI 10.1007/978-3-319-52181-7_8

This flexibility is not for free, but leads to new problems and exacerbates existing ones. One of the problems that regain in importance is the problem of application *portability*. Portability, the ability to move software among different runtime environments without having to rewrite it partly or fully [2, 3], is the prerequisite for benefiting from current trends and a primary enabler for the evolution of process-aware information systems (PAIS). There is no use in being able to provision a new runtime environment, if the target application cannot be adjusted to run in that particular new environment. Application portability is a central characteristic of software quality [4] and is part of various software quality models, e.g., [5–9]. Especially since the arrival of cloud-based applications, work on application portability, e.g., [10, 11], is gaining momentum. This is also demonstrated by recent standardization initiatives, such as the *Topology and Orchestration Specification for Cloud Applications* [12].

Process-aware applications [13] are in a premier position to address these problems and have the potential to better cope with them than traditional applications, as for instance demonstrated by the recent advent of cloud-based process management systems [14]. This results from the fact that the abstraction from the execution platform, i.e., the runtime *engine*, is a fundamental concept in the architecture of process-aware information systems [15]. If a process is developed in a format that is independent of a concrete engine, it should be easily portable to any engine that consumes this format. Major international process standards, such as the *Business Process Model and Notation* (BPMN) [16], the *Web Services Business Process Execution Language* (BPEL) [17], or the *XML Process Definition Language* (XPDL) [18], name the portability of processes as an important goal and provide platform-independent serialization formats that can solve the problem of portability. Theoretically, if a process is implemented in conformance to a standard, it should be executable on and portable to any implementation of the standard. This abstraction can be leveraged in a cloud-based process execution scenario, which is depicted in Fig. 8.1. Instead of directly deploying a process to a specific engine, a broker agent can automatically select the most suitable one based on metric data and deploy it there. In [19], this example is explained in more detail for the case of BPEL processes and engines. Additionally, the agent can adapt and improve the process before deployment.

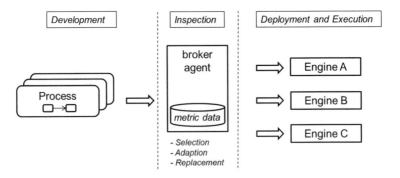

Fig. 8.1 Engine selection with the help of a broker agent

Despite this extraordinary starting position, process-aware applications today are still limited with respect to their portability. The implementations of international standards rarely support the complete specification, but omit certain parts or differ in the interpretation of the standard [20–22]. As a consequence, even processes that are compliant to a standard might not be portable to any process engine for that standard, if they use features that no engine supports.[1] In other words, it is not possible to depend on a standard alone for achieving process portability. This is an obstacle to PAIS evolution and the application scenario from Fig. 8.1. In this situation, agile techniques for software quality improvement, such as continuous inspection [24, 25], can be applied as a remedy [24, 26–29]. The prerequisite for the application of such techniques is that portability issues can be detected and quantified. This quantification is the central topic of the current chapter. Work on measuring the portability of applications, and specifically of process-aware ones, is rather scarce. For this reason, we work on a metrics suite for quantifying the portability of and detecting portability issues in process-aware applications, based on recent software quality standards [5]. Such software metrics could be leveraged to support an easier evolution of processes and execution engines. For instance, when a new version of an engine becomes available, metric data can be used to confirm if existing processes can automatically be ported to said engine. As explained in Fig. 8.1 and outlined in [19], with the help of metric data, an agent could automatically perform the selection of an appropriate engine for a given process. The agent could analyze the process and, based on the constructs it uses, determine on which engines the process can be executed. From this set of possible engines, the most suitable one can be selected based on further quality data or heuristics, such as the ease of its installation. If no suitable engine for a process is found, metrics and inspection methods can be used to identify possibilities for adaptions of the nonportable parts of the process or even replacement candidates for the process as a whole. To sum up, support with software metrics has the potential to reduce the manual effort required for porting processes and to ease the evolution of process-aware information systems.

This chapter is intended as an overview on the topic of portability for process-aware information systems. In Sect. 8.2.1, we explain in detail why problems of portability are not solved by today's standards as discovered in several case studies. We then go on and describe in Sects. 8.2.2 and 8.2.3 how these problems can at least be relaxed through techniques of software measurement and continuous inspection. Thereafter, we give an overview on our current work on software metrics for portability in Sect. 8.3, along with examples explaining how an intelligent PAIS can make use of metric data. The chapter is based on several publications [3, 20, 21, 30, 31] which we extend and synthesize into a unified context.

---

[1]This problem is evident if processes are not implemented manually for a particular engine, but derived automatically, for instance, in a model-driven mapping [23].

8.2   How Software Measurement Can Help

The measurement of software quality is almost as old as the discipline of software engineering itself [32, 33]. The quantification of the quality of a software product can lead to an improvement of its quality [26].

In the following subsection, we show why the problem of process portability is not already solved by the international standards and specifications that exist. Thereafter, in Sect. 8.2.2, we discuss how the usage of agile techniques such as continuous inspection can help to improve this situation, given a mechanism for the detection of portability issues and a quantification of portability is available. This leads to a description of several frameworks that form the basis for this quantification in Sect. 8.2.3, in particular the ISO/IEC 25010 standard for software quality [5]. The next subsection is partly based on [21].

8.2.1   Standards Are Not Enough

Several standards and notations for building process-aware systems exist today, e.g., [16–18]. It is a popular conception that all that has to be done to achieve portability of process code is to write it in one of these notations. The availability of a standardized serialization format for processes is enough to silence most arguments relating to portability. This is an obvious advantage for the implementers of a standard and vendors of tooling. In the absence of certification authorities, it is easy to claim support for a given standard.

However, the reality of implementing process-aware applications looks different. Process specifications are complex and might contain ambiguities, as discussed for particular specifications in [22, 34]. Moreover, they often have a large set of language elements. For instance, the BPMN specification lists 63 different types of events in Sect. 10.4 [16]. As a result, the implementers of such a language often just implement a subset of the language or implement some language features in a way that differs from the original language specification. Only the elements of the language that are contained in the overlap of these subsets are truly portable. Other elements are only portable to a limited degree, as depicted in Fig. 8.2. This implies that a practical porting of a process is often not feasible, despite the fact that multiple implementations claim to support the language the process is implemented in, and portability should therefore theoretically be a given fact.

To shed light on this situation, we performed two studies [20, 21], in which we benchmarked and analyzed a variety of process engines to see how well they support the language they claim to implement. We chose BPEL [17] for these studies. This language has been conceived for service orchestration and has received tremendous interest in academia and industry since the publication of its final version in 2007. Today, this interest is somewhat in decline in favor of other languages, but a variety of runtimes for BPEL have been built and are used in production today. Porting BPEL

Fig. 8.2 Different subsets of supported language elements by different engines

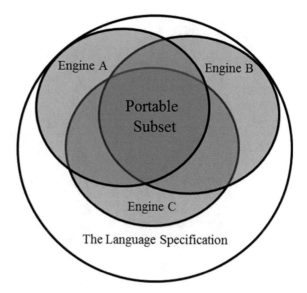

processes is a real problem practitioners face. This makes an analysis of the runtimes of this language worthwhile.

To see how well process engines support the language, we implemented a conformance benchmarking tool [35], called *betsy*.[2] This tool provides a conformance test suite which covers all language elements of BPEL with more than 130 test cases, in the form of standard-conformant BPEL processes. Additionally, we also provide tests for the functional correctness of implementations of workflow control-flow patterns. Moreover, betsy can be used to automatically manage (download, install, start, and stop) several open source and commercial engines and execute the tests in an isolated fashion [36]. We derived conformance tests from the normative parts of the BPEL specification. The test suite is subdivided into three groups, namely, *basic activities*, *scopes*, and *structured activities*, resembling the structure of the specification listed in Sects. 10–12 [17]. The various configurations of the BPEL activities of each group form the basis of the test cases, including all BPEL faults. Hence, every test case of the standard conformance test suite asserts the support of a specific BPEL feature. Firstly, every test case consists of a test definition, being the BPEL process definition and its dependencies (WSDL definitions, XML Schemas, etc.). The second part of a test is the test case configuration, being the specification of the input data and assertions on the result. The aim of every test is to check the conformance of a single language feature in isolation. During a full test run, our tool automatically converts the engine independent test specifications to engine specific test cases and creates required deployment descriptors and deployment archives. Next, these archives are deployed to the corresponding engines and the test case configurations

---

[2]More information on the tool and a description of its usage is available at the project homepage located at https://github.com/uniba-dsg/betsy.

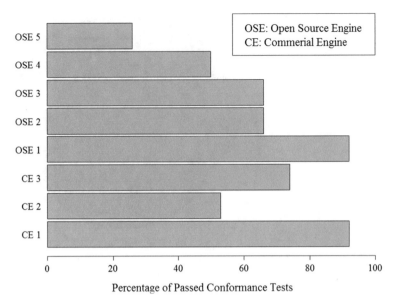

Fig. 8.3 Standard conformance of process engines—the figure displays the percentage of success-fully passed tests per engine as discussed in [21]

are executed. At first, every test case configuration checks successful deployment and, thereafter, performs the different test steps. These test steps send messages and assert the correctness of the responses by means of return values or expected SOAP faults. When all test cases have been executed, HTML and CSV reports are created from the test results. The complete test procedure is quite complex and takes several hours to execute, due to the complexity of the installation of several of the engines. In the end, a comprehensive overview over the support for the BPEL specification is the result. For a more detailed explanation of the testing procedure, we refer the interested reader to [20, 21, 35].

Figure 8.3 provides a rough overview of the state of BPEL support and the most important results from [21]. It shows the percentage of passed tests of our test set for a variety of engines, both of commercial and of open source origin. Two main problems can be read from the plots: Firstly, no engine passes all the tests, hence no engine implements the complete specification. Secondly, the variances in the amount of successful tests is high for the different engines. In more detail, proprietary engines successfully pass between 53 and 92% of the conformance tests. For open source engines, these numbers vary from 26 to 92%. On average, proprietary engines pass 73% of the conformance test suite, whereas the open source engines only achieve 62%. In total, proprietary engines provide a significantly higher degree of support, although the difference balances if we only consider mature open source engines. All in all, the data demonstrate that the porting of processes among these engines will be difficult, due to different degrees of language support.

A similar investigation is currently ongoing for the BPMN language [16] and we extend betsy for benchmarking BPMN engines. Although we cannot yet unveil the detailed results, we can say that the situation is very similar if not even more critical for this language. In any case, the results presented here seem to be representative for the situation we face in practice, regardless of what process language we consider. They demonstrate, that the availability of process standards alone does not guarantee the portability of processes. Implementations will always deviate from a standard for a variety of reasons, such as economic constraints, issues in the specification, or software errors. The best we can do is to adapt to this situation and to try to build portable processes nonetheless. This is necessary to better cope with today's highly dynamic environments and to support the evolution of process-aware information systems. Intelligent support through software measurement and agile techniques such as continuous inspection, which are detailed in the following section, can help in this task.

8.2.2 Quality Improvement with Continuous Inspection

Continuous inspection is a term for the convergence of two quality assurance techniques, *software inspection* [37] and *continuous integration and delivery* [24, 38]. Continuous inspection refers to the constant and automated inspection of a software product for every source code commit to enhance its quality [25]. In this section, we explain how this combination works and why it has the potential for the improvement of process portability.

Software inspections, pioneered by Fagan [37], have a long tradition in software engineering [39, 40]. Ordinarily, these are manual tasks performed as a quality assurance technique next to other techniques such as unit testing. Essentially, an inspection is a review process where a team of reviewers individually scrutinize a software product according to a predefined set of criteria. The reviewers try to verify if the product meets its specifications and has a sufficient level of quality [41]. Afterwards, the reviewers gather in a meeting and produce a list of defects that can be handed to the authors of the software to fix these issues. Obviously, this process requires a lot of communication and is therefore expensive to perform, especially in a repeated fashion. For these reasons, inspection tools have emerged during the last decade. These tools are static code analyzers that automatically highlight potential issues in code [42]. The benefit of their usage is that an inspection can normally be performed within mere seconds and repeatedly. Of course, such a tool might not find all issues or detect false positives, but it offers unprecedented advantages in terms of efficiency. Software inspections need not necessarily be limited to the detection of potential issues, but are also a good occasion to compute quality metrics to see if the software meets predefined quality criteria.

Inspection tools can be used to much benefit, when combined with *continuous integration* (CI) [24]. CI is a term that emerged in the context of agile software development methods. It belongs to the concept of *continuous delivery* [38] which

Fig. 8.4 Continuous
inspection cycle adapted
from [25]

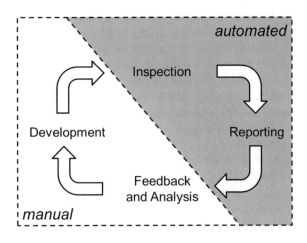

is specified in the first principle of the agile manifesto [43]. CI is a technique applied during software development that refers to the frequent integration of all parts of a program and the validation that they do work together properly. In practice, a continuous integration server is set up and configured to build the program and run the complete set of tests available every time a commit is made to the version control system [24]. This allows getting immediate feedback for the complete team in every stage of development and offers a variety of benefits, as described on pages 17–22 of [38], or 29–32 of [24]. For instance, defects that are newly introduced into the code can be detected immediately and fixed at a point in time where the cost of correction is relatively low. CI has been embraced in practice and a variety of CI servers and tools are available.

With the proper tools, both of these techniques, software inspection and continuous integration, can be combined to continuous inspection [25]. The idea depicted in the feedback cycle in Fig. 8.4 is to not just run tests for every commit to the version control system, but also to inspect the code automatically with available inspection software.

That way, possible defects that are not captured in the tests can be discovered and fixed just as quickly. It might even be possible to detect issues in the code that have not yet turned into concrete defects, but are likely to do so in the future. Moreover, this is an opportunity to compute software metrics and compare them to previously configured thresholds [24, 38]. This way, it can be directly perceived if software quality deteriorates and counter measures can be taken before the deterioration turns into software errors. What is more, through this feedback, developers learn which patterns of code tend to reduce quality and which ones tend to improve quality and are encouraged to produce code of higher quality,[3] as described on pages 137–140 of [38].

---

[3]This effect of the influence of the measurement on the persons being measured is known as the *Hawthorne effect*. Although it normally is disruptive for experiments, it can also be used to train developers in the fashion described in the text.

The focus of this chapter is portability. So, applying continuous inspection to this context, this section can be summarized in the following: Given inspection methods and tooling are available for detecting portability issues in code, these methods and tools can be used in today's ubiquitous CI environments through continuous inspection and thus have the potential to lead to higher code quality. In the terms of this chapter, higher code quality refers to more portable code. Here, we present methods and tools for measuring the portability of process-aware information systems. The computation of the metrics we propose is implemented in an open source and freely available inspection and issue detection library, essentially a static analyzer, the *prope* tool.[4] That way, the metrics can be integrated into the continuous inspection cycle. Referring to the motivating scenario from the introduction depicted in Fig. 8.1, this computation can be leveraged by a broker agent. For every source code commit, a process can be inspected automatically by prope. The inspection results in a list of metric values and issues found in the process, available as CSV and XML reports. An agent can process these reports and use the data to select a proper engine for execution. For instance, engines that do not support one or more activities which the process uses can be immediately excluded from the selection.

8.2.3 Portability and the ISO/IEC 25010 Quality Model

To use a technique such as continuous inspection, it is necessary to be able to capture software quality in the first place. This is where software quality models, e.g., [5–9], come into play. An abundance of quality models has been developed during the last decades.[5] Nevertheless, older quality models, especially by Boehm et al. [6], Gilb [7], or McCall [8], are still very influential. Such models typically define a hierarchy of *quality characteristics* of software, sometimes also called *quality attributes* (cf. Sect. 3.6 of [7]). Each quality characteristic describes a certain major aspect of software quality. Examples are characteristics such as *performance efficiency, usability,* or *portability* [5]. It is often useful to divide these high-level characteristics into a number of *subcharacteristics* that focus on more specific aspects.[6]

For instance, performance efficiency can be divided into the subcharacteristics of *time behavior, resource utilization,* and *capacity* [5]. The main difference between different quality models [5–9] lies in what quality characteristics and subcharacteristics they define and how many layers of characteristics they use.

The quality model used here stems from the ISO/IEC series of quality standards, 9126 [9] and 25010 [5]. The characteristics defined by the model *"are widely accepted*

---

[4]Prope stands for *PROcess-aware information systems Portability mEtrics suite.* For more information on this tool and instructions on how to use it, please visit the project page located at http://uniba-dsg.github.io/prope/.

[5]The amount of quality models has also led researchers to build models of models, such as the systemic quality model presented in [4].

[6]This is called the *attribute hierarchy principle,* defined on page 135 of [7].

both by industrial experts and academic researchers" [44, p. 68] and often cited
(just to mention a few, cf. [4, 26, 45, 46]). This series of standards is very prevalent,
since it has been derived from and synthesizes other well-known quality models,
e.g., [6–8], and, moreover, has received widespread acceptance in industry due to
the standardization process. Software vendors pay a considerable amount of money
to obtain an ISO certification.

The ISO/IEC 9126 series [9] is most renowned, but is currently being revised
in the context of the ISO/IEC 25010 series [5]. This series is titled "*Systems and
software engineering—Systems and software Quality Requirements and Evaluation
(SQuaRE)—System* and software quality models" and will completely supersede the
9126 series when it is finished. The quality model from [5] is depicted in Fig. 8.5. It
defines eight top-level quality characteristics and one of these is portability, the focus
of this chapter. Portability has three subcharacteristics, *adaptability*, *replaceability*,
and *installability*. The reasoning behind this structuring can be expressed through a
sequence of decisions made when porting an application, as depicted in Fig. 8.6. If
a software product, in our case a process, needs to be ported, the starting point is to
check if the process can be *directly ported* in its current form. If this is not the case,
there are basically two options:

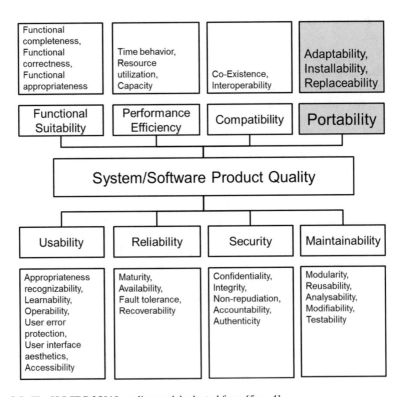

Fig. 8.5 The ISO/IEC 25010 quality model adapted from [5, p. 4]

Fig. 8.6 Portability and its
subcharacterisitics

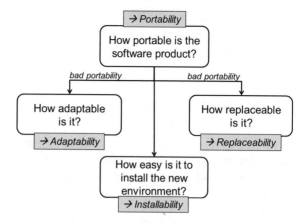

1. The nonportable parts of the process can be adapted for the new environment. The ease of this depends on the *adaptability* of the process.
2. The process can be *replaced* as a whole by an alternatively available process that runs on the new platform. This depends on whether a suitable alternative is available.

In any case, the new runtime environment for the process has to be *installed*. [5] clarifies that portability and its subcharacteristics have significant influence on the working life of the maintainers and operators of a system. This is especially true for the installability of the system, since the operators of an organization will be the ones who have to perform the installation. As a consequence, an improvement of portability leads to an improvement of the working life of the operators. Enabling the quantification of these characteristics during development provides developers with feedback that allows them to develop software that is better to operate. This integration of *development and operations* is the central goal of the *DevOps* movement [47]. DevOps is a term for another agile practice that aims at improving IT performance and is strongly related to continuous integration and delivery. It is currently receiving widespread attention and is increasingly adopted in practice [48].

Each of the quality characteristics should be quantified to allow for meaningful decisions. The enabling of this quantification is our long-term goal and the content of the remainder of this chapter. At the time of writing, the specification that is to contain concrete metrics [49] is still under development and not yet open to public scrutiny. Moreover, the metrics from preceding versions of the ISO/IEC quality standards [50, 51] are rather general and often based on the observation of human behavior, which is, arguably, not the decisive factor when it comes to the executability of processes on different engines. In previous work, we proposed and validated measurement frameworks for the direct portability of processes [3] and their installability [31]. We are currently working on a similar study for adaptability [30], whereas replaceability still remains open. In the following sections, we give an overview over the proposed frameworks for direct portability and installability with a brief description of the metrics, outline the state of our work for adaptability and provide an outlook on replaceability.

8.3 Metrics for Process Portability

In the following, we discuss how the different characteristics related to portability, *direct portability*, *adaptability*, *replaceability*, and *installability* can be measured for process-aware information systems. We present current metrics, as far as they are available yet. The metric definitions are taken from the respective publications [3, 30, 31].

Our focus lies on the presentation and description of the metrics. Their validation and evaluation is a critical factor, but, due to page constraints, we cannot fully lay out a complete validation and have to refer to the literature [3, 31]. There, we validate the metrics we propose using two theoretical validation frameworks related to *measurement theory* [52] and *construct validity* [53] and complement the validation with an experimental evaluation of process libraries.

8.3.1 Direct Portability

Direct portability is the ability to directly take a piece of software and execute in on another platform without modification. It is a quality characteristic that is typically hard to quantify with reasonable effort [45] and no corresponding metrics for measuring it can be found in the respective ISO/IEC standards [50, 51]. This section is based on [3], where we present an approach for measuring direct portability of processes and evaluate it with BPEL processes.

A practical way for measuring direct portability is by contrasting the complexity of the task of porting a piece of software to the complexity of rewriting it from scratch [45]. To capture this complexity, existing metrics for portability use a lines of code-based calculation. This approach can also be applied to process-based software, but we can improve the accuracy of the measurement by taking domain knowledge of process-aware systems into account. In summary, a portability metric can be based on the following equation:

$$M_{port}(p) = 1 - C_{port}(p)/C_{new}(p) \tag{8.1}$$

$M_{port}(p)$ is a metric that quantifies the degree of portability for a process p. A process can be characterized as a tuple of three sets, $< E, A, S >$, where E is the set of elements of the process, A the set of activities, and S the set of communication activities. Activities are also elements, so $A \subset E$ and also $S \subset A$ applies. $C_{port}(p)$ is the complexity of modifying the process so that it can run on another platform.

$C_{new}(p)$ is the complexity of rewriting it completely for a new platform. Equation (8.1) is based on the assumption that the complexity of a rewrite is always at least as high as the complexity of modification. This implies that the metric value ranges in the interval of zero and one, where zero indicates no portability and one full portability. The difficulty in this equation is how to actually determine the complexity. The different metrics presented here propose different ways of calculating these values.

As described in Sect. 8.2.1, the direct portability of a process is strongly tailored to the engines that exist for the language the process is written in. Only process elements that are supported by a majority, or all, engines can be considered to be portable. As a consequence, the measurement of process portability should take the engines for said processes into account, and not be based on a theoretical consideration of the problem only. If all available engines support all the existing language elements in the same manner with respect to semantics, then any executable process will be portable to any engine and there are no portability issues. Section 8.2.1 demonstrates that the situation is rather different in practice. Each engine typically supports a specific language subset, as depicted in Fig. 8.2, causing portability issues. On the one hand, there is a basic subset of the total language that is fully portable. On the other hand, several language elements are only portable in certain configurations or are limited to a subset of engines. The more engines support a language element, the more portable it can be considered.

To take this into account, we calculate a *degree of severity*, referred to as D_{ta}, with respect to portability for each language element and its configuration. This degree can be identified by the number of engines that do not support an element. The smaller the amount of engines supporting a language element, the harder it will be to port a process that uses this element. We can precisely determine this amount, by considering the engine benchmarks described in Sect. 8.2.1. The benchmarks list for every language element, whether it is supported by a given engine. This enables us to statically check processes for elements that are not supported by all engines, as discovered by the benchmark. The portability metrics we propose describe different aggregations of the support for every language element used in a process to a portability value for the overall process. We consider a high-level view, typical for classical portability metrics, a process-oriented view and a communication-oriented view.

In combination, these metrics form a comprehensive framework for quantifying portability, which is depicted in Fig. 8.7. Although all the metrics focus on direct portability, the scope is reduced for each metric to a more limited, but also more critical part of a process. Simply put, the *basic portability* metric takes all process elements equally into account. The *weighted portability* metric also considers all process elements, but includes engine support data in the computation. *Activity portability* only takes control-flow related aspects, such as activities, gateways, or events, into account. Finally, *service communication* portability only considers activities related to message sending and reception.

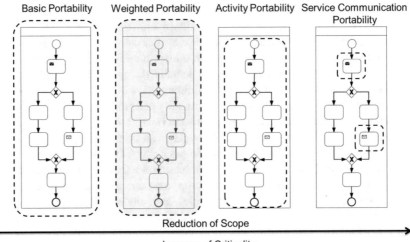

Fig. 8.7 Framework for measuring direct portability—dashed rectangles mark the scope of a metric

Basic Portability A universally applicable way of calculating $C_{port}(p)$ and $C_{new}(p)$, denoted as *basic* portability metric M_{basic}, is to consider the lines of code that have to be rewritten for porting the software (as indicated in [6, 7]). If it is to be redeveloped from scratch, all lines will have to be rewritten, so $C_{new}(p)$ amounts to the total lines of code of the program. In our case, lines of code correspond to process elements, so $C_{new}(p)$ refers to the total amount of elements in a process, denoted as N_{el} being the cardinality of set E. $C_{port}(p)$ in turn amounts to the elements from E that have to be rewritten when porting it, i.e., the number of elements from E for which problems could be detected. As the number of elements that have to be rewritten for porting cannot be larger than the number of elements that do actually exist, $C_{port}(p) \leq C_{new}(p)$ always applies. In the most extreme case, where all elements are nonportable, $C_{port}(p)$ will be equal to $C_{new}(p)$ and consequently $M_{basic}(p) = 0$, indicating no portability at all. The metric is undefined for an empty program, where $C_{new}(p) = 0$.

Weighted Portability M_{basic} transfers the classical abstract portability metric to the area of process languages. However, it does not make full use of the empirical data at hand. To be precise, it only confronts the amount of fully portable elements of a process to all of them. Using the degree of severity D_{ta}, described at the beginning of this section, it is possible to fine-tune this observation, resulting in a more accurate metric value. This is the principle underlying this and the following metrics.

For the *weighted elements* portability metric, the complexity of rewriting a process C_{new} is defined as $C_{new}(p) = N_{el} * N_{engines}$. It is identical to the amount of elements N_{el} (as in the basic portability metric) multiplied with the number of engines under consideration $N_{engines}$. Effectively, every element is treated as if it is unsupported by any engine and has to be rewritten when being ported, resembling the worst case. The complexity of porting C_{port} is defined as follows:

$$C_{port}(p) = \sum_{i=1}^{N_{el}} C_{el}(el_i) \qquad (8.2)$$

The complexity of porting C_{port} of a process p is the sum of the element complexity C_{el} for each element el_i from E. The element complexity C_{el} for an element el_i of process p refers to the most severe portability issue that can be detected for the element.[7] It corresponds to the degree of severity as defined above, the number of engines not supporting the element in its current configuration. The more engines that support the feature, the less the complexity of porting it will be. Similarly, the fewer the number of engines, the higher the complexity. Summarizing the above discussion, the weighted elements metric M_{elem} is calculated as follows: $M_{elem}(p) = 1 - \sum_{i=1}^{N_{el}} C_{el}(el_i)/(N_{el} * N_{engines})$.

Activity Portability The most central building block of process languages in general are activities. Activities are typically basic atomic steps of computation. For process complexity measures [54, 55], activities and the transitions among them are the dominant factor. Apart from activities, processes include a variety of other elements such as, for instance, variable definitions. Considering the conceptual importance of activities, it can be expected that the impact of using problematic activities on portability is critical. Having to alter the flow of control for porting a process affects its behavior which is not desirable.

An activity-oriented view on portability is provided by the *activity* portability metric M_{act} as a variation of the weighted elements metric. Here, instead of elements, we only consider activities and problematic configurations thereof (i.e., the elements of set A) when computing portability. Issues that cannot be linked to a specific activity, as for example process-level *import* statements or variable definitions, are omitted in the consideration of this metric. For M_{act}, C_{new} changes to $C_{new}(p) = N_a * N_{engines}$ where N_a denotes the total amount of activities, the cardinality of A, in the process definition. C_{port} changes to $C_{port}(p) = \sum_{i=1}^{N_a} C_{el}(a_i)$. This means that only the element complexity C_{el} of the activities in p is considered.

Service Communication Portability Communication and composition relations among services and processes are a decisive factor for service-oriented and process-aware systems and metrics for such systems center on these properties [56]. Communication relationships describe the *observable behavior* of a process; that is, the messages it sends and receives. The distinction between the description of observable and internal behavior is the discriminating factor for different types of process models [57]. Message sending and reception is performed using specific activities. In terms of portability, these activities are most critical. Single elements and perhaps

---

[7]This is a simplified description of element complexity. For an extensive discussion and formal definition of this function, please refer to [3].

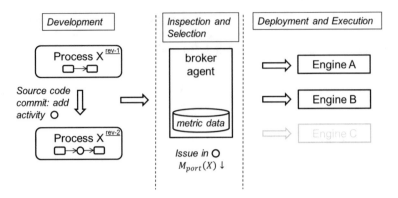

Fig. 8.8 Engine selection based on portability data

even the control-flow structure of a process may be changed for porting in a way that does not affect the observable behavior. However, this is unlikely if the activities that have to be changed concern communication. In this case, these activities directly affect the observable behavior of the process. Changing them (to enable portability) and consequently changing the observable behavior influences other systems that interact with the process, and this way of change propagation is highly undesirable.

The service communication portability metric M_{serv} allows focusing on the impact of communication related activities on portability. For this metric, the calculation of C_{new} and C_{port} is changed to include only the activities relating to service interaction (i.e., the elements of set S), that is: $C_{new}(p) = N_s * N_{engines}$ and $C_{port}(p) = \sum_{i=1}^{N_s} C_{el}(s_i)$. Effectively, this is an extension of M_{act} that focuses solely on activities for service interaction. N_{serv} refers to the total amount of activities for service interaction, the cardinality of S. C_{port} is limited to only consider the element cost of these activities.

In Summary The metrics presented in this section and depicted in Fig. 8.7 try to capture the *direct portability* of a process. For more details and a validation and evaluation of the metrics, please refer to [3]. Each of the metrics corresponds to a reduction in scope and an increase in criticality. In combination, they allow for a comprehensive view of the portability of a process. If evaluated in the continuous inspection cycle, as described in Fig. 8.4, they can be used to immediately detect a deterioration of the direct portability of a process under development. Figure 8.8 considers the scenario of automatic engine selection presented in the introduction. A new activity is inserted into process X and, during inspection, an issue is detected in the configuration of this activity, resulting in a decrease of its portability. The metric values for the process and the issue are reported by the inspection tool, prope, and this information can be utilized by a broker agent. If, for instance, the activity is known to be unsupported by engine C, this engine can be excluded from the selection by the broker agent.

8.3.2 *Installability*

The ISO/IEC SQuaRE model defines installability as the *"degree of effectiveness and efficiency with which a product or system can be successfully installed and/or uninstalled in a specified environment"* [5, p. 15]. When it comes to process-aware information systems, two components are of interest with respect to installability: The process itself and its runtime engine.

Both of these components influence the installability of the complete system and should therefore be taken into account. Figure 8.9 outlines the model we use for measuring installability. In [31], on which this section is based, we divide the quality characteristic *installability* into the subcharacteristics *engine installability* and *deployability*.

Each of the subcharacteristics can be measured by a set of direct and aggregated metrics. Direct metrics can be computed directly from source code artifacts or log files, whereas aggregated metrics are formed by the combination of direct metrics. The metrics *ease of setup retry (ESR)* and *installation effort (IE)* stem from the ISO/IEC standards [50, 51]. We extended installation effort to also consider *average installation time (AIT)* and not only the *number of distinct steps (NDS)* required for the installation. When it comes to deployability, no corresponding metrics are available in [50, 51], so we develop new ones. These consist of *deployment effort (DE)*, which considers *deployment descriptor sizes (DDS)* and the *effort of package construction (EPC)*, next to *deployment flexibility (DF)*. The deployability metrics *DDS*, *EPC*, and *DE* are *internal* (i.e., they relate to static properties of the software),

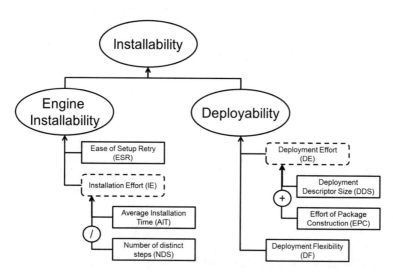

Fig. 8.9 The framework for measuring installability adapted from [31]. Ellipses denote quality characteristics, rectangles denote direct metrics obtained through code analysis and benchmarking, and rounded dashed rectangles depict aggregated metrics that are computed by the combination of direct metrics using the functions displayed in circles

and the remaining metrics are *external* (i.e., they relate to dynamic properties and can be verified during execution), as specified in Sect. C.3 of [5].

Ease of Setup Retry (*ESR*) This metric, defined in Sect. 8.6.2 of [50], is intended to measure how easy it is to successfully repeat an installation of an engine. It relates the number of successful installations of the same engine e (N_{succ}) to the number of attempted installations in total (N_{total}). That is $ESR(e) = N_{succ}/N_{total}$. [50] refers to manual installations, but the metric is just as applicable to an automated installation process. If this process is completely deterministic, then those numbers will be identical and $ESR(e)$ equal to one. If it is not free of errors, installations may fail, resulting in a lower *ESR* value.

Installation Effort (*IE*) Installation effort provides a notion of the difficulty of the installation process of an engine. Section 8.6.2 of [50] suggests to measure it as the amount of automatable installation steps in relation to the total amount of prescribed steps. As we found in the case study described in [31], the installation of process engines can often be automated fully, but the complexity of this automation and, more importantly, the duration of the installation varies a lot. For that reason, we deviate in the measurement of installation effort from [50] and instead measure it through a combination of two direct metrics: The total number of distinct steps (*NDS*) and the average installation time (*AIT*). The first is identical to the number of steps that need to be automated, and thereby partly corresponds to the metric defined in [50]. *NDS* includes every operation that needs to be performed for the installation, such as the copying of files and creation of directories or changes in the configuration of certain files. This metric can be determined through a *heuristic evaluation* [58], i.e., we can essentially count each step in an installation script. The average installation time can be computed by performing the distinct steps required a suitable amount of times and measuring execution times. *AIT* and *NDS* can be aggregated to a notion of installation effort (*IE*) per installation step:

$$IE(e) = \begin{cases} 0 & \text{if } NDS = 0 \\ \frac{AIT(e)}{NDS(e)} & otherwise \end{cases} \tag{8.3}$$

Note that an installation routine that consists of several simple steps is desirable over a single installation step that takes very long even if the multiple step installation takes longer. The reasoning behind this is that simple and quick installation steps are easier to automate, to repeat in case of a failure, or to adapt to a new environment.

Deployment Flexibility (DF) This metric can be used to quantify the availability of alternatives for achieving the deployment of a process. Deployment normally consists of the execution of a single engine operation provided with all artifacts needed for execution. Nevertheless, deployment can take different forms, multiple of which can be supported by an engine. The more options a server supports, the more flexible it is and the easier deployment can be achieved. We capture this in the metric *deployment flexibility* (*DF*), which corresponds to the number of options available. The intention of the metric is to adapt *installation flexibility* from Sect. 8.6.2 of [50] to this context. Typically, three different options are available:

1. Hot deployment, i.e., a copy operation of a deployment archive into a specific directory,
2. the invocation of a deployment script, or web service,
3. a manual user operation using a GUI or web interface.

Deployment Effort (DE) Being similar to installation effort, this metric can be used as an overview of the complexity of the preparation of a process for its deployment. Deploying a process normally requires its packaging and the construction of one or more deployment descriptors. The construction of these descriptors may be partly automated or aided by graphical wizards, but in the end it is configuration effort that can take a significant amount of time to get right. The more complex the packaging and the more extensive the descriptors, the harder it is to deploy a process on a specific engine. We capture packaging with the metric *effort of package construction (EPC)* and deployment descriptors with the metric *deployment descriptor size (DDS)*. The effort of package construction can be measured by counting each part of a prescribed folder structure that needs to be built and compression operations that need to be performed to construct the prescribed deployable executable: $EPC(process) = N_{fc} + N_{dc} + N_{co}$. N_{fc} refers to the amount of folder creations, N_{dc} to the amount of descriptors, and N_{co} to the amount of compression operations required. The deployment descriptor size *DDS* for a process corresponds to the added size of all descriptor files, $\{dd_1, \cdots, dd_{N_{desc}}\}$, needed: $DDS(process) = \sum_{i=1}^{N_{desc}} size(dd_i)$. For process-aware applications, typically two different types of descriptor files exist in practice: (i) plain text files and (ii) XML configuration files. As plain text files and XML files differ in the ways in which they represent information, different ways of computing their size are needed. For plain text files, a lines of code metric is appropriate. For the descriptors at hand, every nonempty and noncomment line in such files is a key-value pair with a configuration setting, such as a host or port configuration, needed for deployment. We consider each such line, using a *LOC* function. For XML files, the notion of lines is not applicable, but instead information is structured in nested elements and attributes. To compute the size of XML files, we consider the *number of elements and attributes N_{ea}*, including simple content and excluding namespace definitions, which represent an item of information in the same fashion as key-value pairs in plain text files. All in all, the *size* of a descriptor *desc* is defined as follows:

$$size(desc) = \begin{cases} LOC(desc), & \text{if } plain(desc) \\ N_{ea}, & \text{if } xml(desc) \end{cases} \tag{8.4}$$

As a result, the deployment effort can be computed in the following fashion: $DE(process) = DDS(process) + EPC(process)$. The idea here is to capture every factor, independent of its nature, that increases the effort of deploying a process.

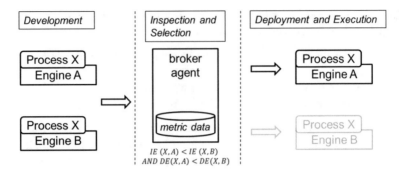

Fig. 8.10 Engine selection based on installability data

In Summary The metrics presented above allow for a quantification of the installability of a process-aware information system. Metrics for engine installability can be used to compare different engines and might help to select the best one. Metrics for deployability can be used to compare different processes with each other and are also suitable for continuous inspection. As demonstrated in [31], certain values for deployability are typical for certain engines. Hence, also deployability metrics can be used as a factor when selecting different engines.

As before, a more detailed formal definition, as well as a validation and evaluation can be found in [31]. Figure 8.10 outlines how this information can be leveraged in the engine selection scenario from the introduction. Prope can be used to compute installability and deployability data of different combinations of a process X and possible engines for X. The agent can compare the metric thresholds of the different systems and select the better one, for instance the one with lesser deployment and installability effort.

8.3.3 Adaptability

When it comes to the quality characteristic of adaptability, we did not yet perform extensive evaluations as in [3, 31]. At the time of writing, work on adaptability is underway, but not yet published. Therefore, we give an outline for this quality characteristic and sketch our preliminary approach for computing metrics, presented in [30], but without defining concrete metrics.

The ISO/IEC quality model defines adaptability as the "*degree to which a product or system can effectively and efficiently be adapted for different or evolving hardware, software or other operational or usage environments*" [5, p. 15]. The definition refers to a manual design-time adaptation. Here, we focus on adaptations to the software environment only. The metrics for adaptability of the existing ISO/IEC quality model [50, 51] are based on counting the number of program functions that seem to be adaptable to different contexts. This number is contrasted with the number of functions that are required to be adapted in the current situation, which is typically all program functions that need to be available in the new environment after porting. By relating these two numbers, one can obtain the percentage of program functions that can be adapted. This provides a basic notion of adaptability for the complete program. However, such a measure is coarse and there is no description of how to actually determine if a function is adaptable or not.

Related studies on adaptability metrics are directed at the architectural layer of a software product and not the concrete source code [59, 60]. There, adaptability is first quantified in a binary or weighted fashion for an atomic element of the respective system, such as a component in the software architecture. These element adaptability scores are then subsequently aggregated using different adaptability indices at different layers of abstraction to arrive at a global value of adaptability for the complete software architecture. This way of computing adaptability should also work when looking at code artifacts and not architectural elements of a program. Here, we focus on executable processes and try to reproduce the adaptability computation in the above sense. Thus, our idea is to quantify adaptability at the level of an atomic process element, such as an activity, and to aggregate this to a global degree for the complete process.

For quantifying adaptability at the level of an atomic process element, we count the number of alternative representations for the functionality provided by the element that result in the same runtime behavior. In every process language, there are typically multiple alternatives for each process element that can result in identical process behavior at runtime. The more alternatives exist for a given process element, the easier it is to replace this element with such an alternative, and hence the more adaptable the resulting code actually is.

A simple example for multiple alternative implementations of the same functionality in BPMN [16] is repetitive execution of a task through a *Loop* marker for the task. Any of the following language constructs can be used to define repetitive execution of a task and hence can be used as an alternative to a *Loop* marker:

1. A combination of an *Exclusive Gateway* and *Sequence Flows*
2. Enclosing the task in a *Loop Sub-Process*
3. Enclosing the task in an *Ad-Hoc Sub-Process*
4. Enclosing the task in an *Event Sub-Process*

It is likely that a BPMN engine will only support a subset of these options. For instance, the Activiti engine,[8] currently does not support normal *Loop* markers (`standardLoopCharacteristics`). It does support the combination of *Exclusive Gateways* and *Sequence Flows*, as well as *Event Sub-Processes*, but no *Loop* or *Ad-Hoc Sub-Processes*. Given that a process with a task that uses a *Loop* marker needs to be ported to the Activiti engine, the code needs be adapted to one of the versions Activiti supports (Fig. 8.11).

---

[8]For more information on this engine, see the Activiti user guide: http://www.activiti.org/userguide/index.html.

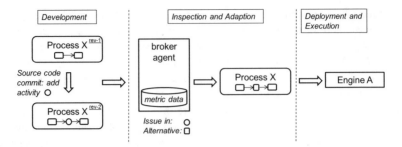

Fig. 8.11 Agent-Aided process adaption

Considering the application scenario from the introduction, the design-time adaptation of a process becomes necessary if no suitable engine for executing it can be found. By inspecting the adaptability of the process, in particular with respect to the elements of the process that hinder portability, it is possible to recommend possible adaptions of these elements. Depending on their nature, the adaptions might even be performed transparently and automatically by the broker agent.

In [30], we propose to capture the number of alternative representations of a process element with its *adaptability score*: $AS(e) =| \{alt_1^e, \ldots, alt_n^e\} |$ This score is equivalent to the cardinality of the set of alternatives $\{alt_1^e, \ldots, alt_n^e\}$ for the element that are available in the language. For the approach to work, such a set of alternatives must be determined for every element of the process language. Based on atomic adaptability scores, a mechanism for aggregating these scores to a global adaptability degree for the complete process is needed. This is necessary to allow for the comparison of different processes in terms of their adaptability. Moreover, the aggregated degree should be normalized with respect to the size of the process, to enable the comparison of processes of different size. A straightforward way of aggregating adaptability scores is the following:

1. Normalize the score for every element.
2. Similar to [59], compute the mean score of all elements in the process.

This leads to the question of how to normalize scores on an atomic level. We propose to divide the score by a reference value. This reference value can be identified by the maximum adaptability score achieved by any of the elements in the language. That way, the most adaptable language element will have a normalized score of one, whereas other elements will have a value between zero and one. This results in the following equation:

$$AM(p) = \overline{(AS(e_1)/R), \ldots, (AS(e_n)/R)} \qquad (8.5)$$

The value of an *adaptability metric AM* of process p, which consists of the elements $e_1, \ldots, e_n$, is equal to the arithmetic mean of the *adaptability scores AS* for every element e divided by the *reference value R*.

For the choice of the reference value, different schemes are possible. The scheme we use here has several advantages with respect to the computation:

1. The resulting metric value always ranges in the interval of $[0, \ldots, 1]$ and thus resembles a percentage value. This scale is easy to understand and interpret, which is critical for the adoption of the metric.
2. The reference value is identical for processes of the same language. Using a reference value that is specific to a concrete process might result in a more meaningful metric value for that process, but it would no longer be directly comparable with different processes. That way, the metric would lose one of its primary purposes.

The division by a reference value is a first proposal for computing an adaptability metric. Alternative ways are possible and we currently test and compare different schemes of computation. We perform an evaluation for BPMN and test the metrics performance for a process library. This way, we hope to find an appropriate way of quantifying adaptability for executable processes. Adaptability metrics computed in this fashion are applicable to continuous inspection.

8.3.4 Replaceability

Also for replaceability, no dedicated metrics have been proposed and tested so far. Hence, we provide a discussion on the nature of the characteristic and present a literature review of existing metrics that could be used for its quantification.

The ISO quality model defines replaceability as the *"degree to which a product can replace another specified software product for the same purpose in the same environment"* [5, p. 15]. This implies that replaceability cannot be evaluated for a single isolated piece of software, but requires a paired combination of two pieces: the currently installed software and a candidate for replacement. In our case, this translates to a currently running process and its replacement candidate.

Similar to the case of adaptability, the evaluation of replaceability for a process becomes necessary, when no suitable engine for executing it can be found. Replacing the process as a whole is an alternative to adapting it, but is only possible if a suitable candidate is available, for instance in a process repository. If this is the case, replaceability metrics can be used to determine the best available replacement. This application scenario is depicted in Fig. 8.12. It might even possible to combine the replacement with adaptability data, to find the most similar process which is the easiest to adapt. This scenario implies that replaceability is typically computed on an ad hoc basis to find an alternative process. In contrast to the other metrics proposed so far, metrics for replaceability are therefore not suitable for continuous inspection. As before, the metrics presented in [50, 51] are solely focused on the observation of user behavior and, hence, not applicable here.

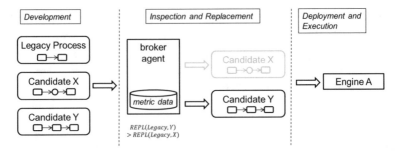

Fig. 8.12 Process selection based on replaceability data

In its core, the question of replaceability is one of *similarity* [61]. Processes are more likely to replace each other if they are highly similar to each other. As a consequence, metrics for process similarity are applicable for evaluating replace-ability. Process similarity is important for a wide array of applications apart from replaceability assessment [61], such as process or service discovery [62–65], compliance assurance [66], pattern support assessment [67], and the facilitation of process change [68], just to mention a few. A correspondingly high amount of similarity metrics has been proposed. These metrics measure similarity among processes in terms of *labels*, *structure*, or *behavior* [65]. Labeling approaches quantify similarity by comparing the names of process elements, structural approaches compare the process graph, and behavioral approaches compare execution traces of processes. The abundance of similarity metrics has led researchers to refrain from the definition of new metrics and perform comparative studies on metrics performance instead. A good example of such a study is [69]. In this study, Becker and Laue perform a review of existing metrics, classify them according to the area of application they are directed at and compute metric values for a set of synthetic process models. Based on this computation, they provide suggestions on the quality and appropriateness of the different metrics. For the application of measuring the conformance of executable processes to each other, which is the problem we face here, the authors recommend measures that compute similarity based on the dependencies among the activities of the process graph. They evaluate metrics based on *dependency graphs* [70] and their improvement in *TAR-similarity* [71], *Casual Behavioral Profiles* [72], *Casual Footprints* [64], and the *String Edit Distance of Sets of Traces* [62]. They find casual footprints to be computationally inefficient and criticize the edit distance of sets of traces, TAR-similarity and dependency graphs for considering direct precedence relationships among activities only, and casual behavioral profiles for being unable to handle OR-splits in process models. Here, an additional comparison of these mea-sures with replaceability in mind would be useful.

8.4 Related Work

We tried to discuss approaches related to the work presented here throughout each section. However, three areas should be examined closer:

1. The standards assessment from Sect. 8.2.1 relates to unit testing and conformance evaluation approaches for process-aware information systems.
2. Software measurement and inspection is, of course, not the only approach for tackling portability issues caused by the dichotomy of specifications and their implementations.
3. A large body of work on software metrics and measurement in general and on metrics for process-aware information systems in particular exists.

In the following three subsections, we discuss each of these areas and provide hints for further reading.

8.4.1 Work on Process Unit Testing and Conformance Validation

The portability metrics we propose build on a conformance testing framework for process-aware information systems, betsy, which relates to other testing approaches in this area. The unit testing of processes, in particular BPEL processes, has received considerable attention [73]. In this area, the *BPELUnit* project [74] is most widely accepted. The main difference between unit testing approaches and the work discussed here is that the former check the correctness of specific process models, whereas we check the correctness of process engines. In other words, the systems under test in our case are different from the systems under test in unit testing approaches.

Conformance checking in the context of process-aware information systems is generally not understood as the testing of the conformance of an engine to the specification it claims to implement. Instead, it refers to the verification of the behavioral properties of a concrete process as specified by an abstract process model. Examples of approaches using this type of conformance checking are [75–78]. Our tool does not check behavioral conformance of concrete process models to abstract specifications. Instead, it checks the implementation conformance of a middleware to a standard specification.

8.4.2 Approaches for Tackling Portability Issues

Studies that address process portability [79–81] also view ambiguities in an informal standard specification as a major problem. [79] try to tackle this problem for BPEL by providing a formal definition of the specification that refines ambiguous aspects. The formalization is accomplished by a formal language called B*lite*. This language can be compiled to executable process code for a specific engine [79]. [80] takes the same approach, by defining a domain specific language that should make programming easier. This approach of pre-compilation can preempt portability problems, by avoiding language elements that are problematic. However, the user of such an approach needs to learn yet another language besides the target one. Here, we do not try to preempt portability issues, but instead to quantify them.

An alternative approach, taken by [81], is to consider the implementation of a standard in practice for improving the standard specification itself. Problems of ambiguity in the specification can be resolved by adopting the interpretation a majority of implementations use in practice. Although we consider the way engines implement the standard in practice here as well, it is not our intent to refine and change the specification, as in [81]. Instead, we determine which aspects of a process definition, although being standard-conformant, cause portability issues and quantify these issues.

8.4.3 Metrics for Selected Quality Characteristics

There is a large body of work on measurement and metrics for process-aware information systems. An overview of the usage of metrics in business process modeling and execution can be found in [82]. General quality metrics for process models build upon classical object-oriented metrics [54], relate to the static complexity of the model during design-time, or the dynamic complexity of the program during run-time [55].

Quality characteristics of processes can often be interpreted, and hence measured, in different ways. In our case, this particularly applies for the quality characteristics of installability, adaptability, and replaceability.[9]

Installability, for instance, can also be viewed as the question whether a set of applications can be installed next to each other on the same machine [83]. Component-, or package-based software systems, such as most Linux distributions, are built from package repositories. Software that is installed into the system might require several other packages in particular versions to be installed as well. These package versions can conflict with the versions required by other software, resulting in a failure of the installation. This contrasts the ISO definition of installability, which we build on here. Nevertheless, the contrasting definition of installability is also covered in the ISO quality model, although it is denoted as a different quality characteristic, *co-existence* [5], there.

Deployability of an application can also be considered as the complexity of its deployment into a network of computers. Here, the complexity of deployment relates to the amount of nodes in the network on which an application has to be deployed to

---

[9]As replaceability and its relation to similarity has been discussed as part of the literature review in Sect. 8.3.4, we omit the repetition of this discussion here.

function properly [84]. In this definition, the complexity of deploying the application on a single host is not considered. Our point of view on deployability is different here. We do not consider the network-wide deployment of an application, but instead the complexity of deploying it on a single host. This view is more fine-grained, but orthogonal to a network-wide deployment and our framework could be combined with such an approach.

Finally, adaptability is used in a different meaning in several other subject areas. In autonomous systems, adaptability refers to the ability of the system to automatically cope with changing situations, such as an increased load, at run-time [60]. Another definition of adaptability can be found in adapter synthesis. There, adaptability refers to whether an adapter for a pair of services can be created [85].

8.5 Summary and Conclusion

In this chapter, we provided an overview on the topic of process portability. We first demonstrated that a problem of process portability exists for today's process-aware information systems. Thereafter, we described how the problem can be tackled through methods of software measurement and continuous inspection, given metrics for portability are available. Following this, we elaborated on our measurement framework and proposed metrics for quantifying portability, with the subcharacteristics of direct portability, installability, adaptability, and replaceability. Though still under development, this framework provides a holistic quantification of portability and, in combination with measurement tools, has the potential to contribute to process portability in the long term. For instance, a process-aware information system can leverage measurement and inspection tools to intelligently control the deployment of a process on a suitable engine.

Several directions of future work follow. The metrics to be used for quantifying the subcharacteristics of adaptability and replaceability still remain open and suitable ones need to be determined. Further testing and refinement of existing metrics would also be useful. Especially in the areas of direct portability and installability only little work is available and the definition of more metrics and their comparison would be valuable. Moreover, it would be helpful to aggregate metric values for different subcharacteristics to an overall quality indicator for portability. This can be achieved using requirements prioritization methods, such as the analytical hierarchy process [86]. Another area of future work is effort prediction. It would be valuable for practitioners if the effort of porting processes could be estimated upfront. Most effort prediction models estimate effort based on the complexity of the software in terms of lines of code. Our metrics are similar to those approaches, since most of them are based on process elements, a notion similar to lines of code. Therefore, it can be expected that there is a relation between effort and our metric values. Nevertheless, an empirical study would be required to determine this relationship.

References

1. Armbrust, M., Fox, A., Griffith, R., Joseph, A.D., Katz, R.H., Konwinski, A., Lee, G., Patterson, D.A., Rabkin, A., Stoica, I., Zaharia, M.: A view of cloud computing. Commun. ACM **53**(4), 50–58 (2010)
2. Petcu, D., Vasilakos, A.V.: Portability in clouds: approaches and research opportunities. Scalable Comput. Pract. Exp. **15**(3), 251–270 (2014)
3. Lenhard, J., Wirtz, G.: Measuring the portability of service-oriented processes. In: 17th IEEE International Enterprise Distributed Object Computing Conference, Vancouver, Canada, September 2013, pp. 117–126
4. Ortega, M., Pérez, M., Rojas, T.: Construction of a systemic quality model for evaluating a software product. Softw. Qual. J. **11**(3), 219–242 (2003)
5. ISO/IEC: Systems and software engineering—System and software Quality Requirements and Evaluation (SQuaRE)—System and software quality models 25010:2011 (2011)
6. Boehm, B., Brown, J., Lipow, M.: Quantitive evaluation of software quality. In: Proceedings of the 2nd International Conference on Software Engineering, San Francisco, USA, October 1976, pp. 592–605
7. Gilb, T.: Principles of Software Engineering Management. Addison Wesley. ISBN-13: 978-0201192469 (1988)
8. Cavano, J., McCall, J.: A framework for the measurement of software quality. Softw Eng Notes **3**(5), 133–140 (1978)
9. ISO/IEC: Software engineering—Product quality—Part 1: Quality model. 9126-1:2001 (2001)
10. Petcu, D., Macariu, G., Panica, S., Crăciun, C.: Portable cloud applications—from theory to practice. Future Gener. Comput. Syst. **29**(6), 1417–1430 (2013). Elsevier
11. Kolb, S., Wirtz, G.: Towards application portability in platform as a service. In: 8th International Symposium on Service-Oriented System Engineering, UK, Oxford, April 2014, pp. 218–229
12. OASIS: Topology and Orchestration Specification for Cloud Applications Version 1.0, November 2013
13. Dumas, M., van der Aalst, W.M.P., ter Hofstede, A.H.M.: Process-aware information systems: bridging people and software through process technology. Wiley. ISBN: 978-0-471-66306-5 (2005)
14. Hoenisch, P., Schulte, S., Dustdar, S., Venugopal, S.: Self-adaptive resource allocation for elastic process execution. In: IEEE Sixth International Conference on Cloud Computing, Santa Clara, CA, USA, June 2013, pp. 220–227
15. Weske, M.: Business Process Management: Concepts, Languages, Architectures, 2nd edn. Springer, Berlin, (2012) (p. Sect. 3.11: Architecture of Process Execution Environments. ISBN: 978-3642286155)
16. ISO/IEC: ISO/IEC 19510:2013—Information technology - Object Management Group Business Process Model and Notation, November 2013, v2.0.2
17. OASIS: Web Services Business Process Execution Language, April 2007, v2.0,
18. WfMC: Process Definition Interface —XML Process Definition Language, August 2012, v2.2
19. Harrer, S., Lenhard, J., Wirtz, G., van Lessen, T.: Towards uniform BPEL engine management in the cloud. In: Proceedings of the CloudCycle14 Workshop Stuttgart, Germany: Gesellschaft für Informatik e.V. (GI), September 2014, pp. 259–270
20. Harrer, S., Lenhard, J., Wirtz, G.: BPEL Conformance in open source engines. In: 5th IEEE International Conference on Service-Oriented Computing and Applications, Taipei, Taiwan, December 17–19 2012, pp. 237–244
21. Harrer, S., Lenhard, J., Wirtz, G.: Open source versus proprietary software in service-orientation: the case of bpel engines. In: 11th International Conference on Service Oriented Computing, pp. 99–113. Springer, Heidelberg (2013)
22. Gutschier, C., Hoch, R., Kaindl, H., Popp, R.: A Pitfall with BPMN execution. In: Second International Conference on Building and Exploring Web Based Environments, Chamonix, France, April 2014, pp. 7–13

23. Lenhard, J., Wirtz, G.: Detecting portability issues in model-driven BPEL mappings. In: 25th International Conference on Software Engineering and Knowledge Engineering, Boston, Massachusetts, USA, June 2013, pp. 18–21

24. Duvall, P.M., Matyas, S., Glover, A.: Continuous Integration: improving Software Quality and Reducing Risk. Addison-Wesley. ISBN-13: 978-0321336385 (2007)

25. Merson, P., Aguiar, A., Guerra, E., Yoder, J.: Continuous inspection: a pattern for keeping your code healthy and aligned to the architecture. In: 3rd Asian Conference on Pattern Languages of Programs, Tokyo, Japan, March 2014

26. Jones, C.: Software quality: the key to successful software engineering, ch. 9. Software Engineering Best Practices—Lessons from Successful Projects in the Top Companies. McGraw-Hill. ISBN: 978-0-07-162162-5 (2010)

27. Dybå, T., Dingsøyr, T.: Empirical studies of agile software development: a systematic review. Inf. Softw. Technol. **50**(9–10s), 833–859 (2008)

28. Karlström, D., Runeson, P.: Combining agile methods with stage-gate project management. IEEE Softw. **22**(3), 43–49 (2005)

29. Huo, M., Verner, J.M., Zhu, L., Babar, M.A.: Software quality and agile methods. In: 28th International Computer Software and Applications Conference, China, Hong Kong, December 2004, pp.520–525

30. Lenhard, J.: Towards quantifying the adaptability of executable BPMN Processes. In: Proceedings of the 6th Central-European Workshop on Services and their Composition, Potsdam, Germany, February 2014, pp. 34 –41

31. Lenhard, J., Harrer, S., Wirtz, G.: Measuring the installability of service orchestrations using the square method. In: 6th IEEE International Conference on Service-Oriented Computing and Applications, Kauai, Hawaii, USA, 16–18 December 2013, pp. 118–125

32. Halstead, M.H.: Elements of Software Science. Prentice Hall. ISBN-13: 978-0444002051 (1977)

33. McCabe, T.: A complexity measure. IEEE Trans. Softw. Eng. **2**(4), 308–320 (1976)

34. Lapadula, A., Pugliese, R., Tiezzi, F.: A formal account of WS-BPEL. In: Proceedings of the 10th Internation Conference on Coordination Models and Languages, Oslo, Norway, 4–6 June 2008, pp. 199–215

35. Harrer, S., Lenhard, J.: Betsy—a BPEL Engine Test System, University of Bamberg, Bamberger Beiträge zur Wirtschaftsinformatik und Angewandten Informatik, no. 90, July 2012, technical report

36. Harrer, S., Röck, C., Wirtz, G.: Automated and isolated tests for complex middleware products: the case of bpel engines. In: IEEE Seventh International Conference on Software Testing, Verification and Validation Workshops, Cleveland, Ohio, USA, April 2014, pp. 390–398

37. Fagan, M.: Design and code inspections to reduce errors in program development. IBM Syst. J. **15**(3), 182–211 (1976)

38. Humble, J., Farley, D.: Continuous Delivery. Addison-Wesley. ISBN: 0321601912 (2010)

39. Aurum, A., Petersson, H., Wohlin, C.: State-of-the-art: software inspections after 25 years, software: testing. Verification Reliab. **12**(3), 133–154 (2002)

40. Ciolkowski, M., Laitenberger, O., Rombach, H.D., Shull, F., Perry, D.E.: Software inspections, reviews and walkthroughs. In: Proceedings of the 22rd International Conference on Software Engineering, Orlando, Florida, USA, May 2002, pp. 641–642

41. IEEE: IEEE Std 1028-2008, IEEE Standard for a Software Reviews and Audits 2008, revision of IEEE Std 1028-1997, Sect. 6. Inspections

42. Foster, J., Hicks, M., Pugh, W.: Improving Software Quality With Static Analysis. In: 7th ACM SIGPLAN-SIGSOFT Workshop on Program Analysis for Software Tools and Engineering, California, San Diego, June 2007, pp. 83–84

43. Beck, K., Beedle, M., van Bennekum, A., Cockburn, A., Cunningham, W., Fowler, M., Grenning, J., Highsmith, J., Hunt, A., Jeffries, R., Kern, J., Marick, B., Martin, R.C., Mallor, S., Shwaber, K., Sutherland, J.: The Agile Manifesto (2001)

44. Ferenc, R., Hegedűs, P., Gyimóthy, T.: Software product quality models. In: Mens T., Serebrenik A., Cleve A. (eds.) Evolving Software Systems, Springer, Heidelberg (2014). ISBN: 978-3-642-45397-7

45. Glinz, M.: A risk-based, value-oriented approach to quality requirements. IEEE Comput. **25**(8), 34–41 (2008)
46. Behkamal, B., Kahani, M., Akbari, M.K.: Customizing ISO 9126 quality model for evaluation of B2B applications. Inf. Softw. Technol. **51**(3), 599–609 (2009)
47. Loukides, M.: What is DevOps? O'Reilly Media (2012). ISBN: 1-4493-3910-7
48. Puppet Labs: IT Revolution Press, ThoughtWorks, 2014 State of DevOps Report, June 2014
49. ISO/IEC: Systems and software engineering—Systems and software Quality Requirements and Evaluation (SQuaRE)—Measurement of system and software product quality, 25023 (2013)
50. ISO/IEC: Software engineering—Product quality—Part 3: Internal metrics. 9126-3:2003 (2003)
51. ISO/IEC: Software engineering—Product quality—Part 2: External metrics. 9126-2:2003 (2003)
52. Briand, L., Morasca, S., Basily, V.: Property-based software engineering measurement. IEEE Trans. Softw. Eng. **22**(1), 68–86 (1996)
53. Kaner, C., Bond, W.: Software Engineering Metrics: What Do They Measure and How Do We Know? In: 10th International Software Metrics Symposium, Chicago, USA, September 2004, pp. 1–12
54. Vanderfeesten, I., Cardoso, J., Mendling, J., Reijers, H., van der Aalst, W.: Quality metrics for business process models. Future Strategies, May (2007)
55. Cardoso, J.: Business process quality metrics: log-based complexity of workflow patterns. In: On the Move to Meaningful Internet Systems, CoopIS, DOA, ODBASE, GADA, and IS, Springer 2007, pp. 427–434
56. Hofmeister, H., Wirtz, G.: Supporting service-oriented design with metrics. In: Proceedings of the 12th International IEEE Enterprise Distributed Object Computing Conference, Munich, Germany, September 2008, pp. 191–200
57. Peltz, C.: Web services orchestration and choreography. IEEE Comput. **36**(10), 46–52 (2003)
58. Nielsen, J.: Usability Inspection Methods. Wiley, New York. ISBN: 978-0471018773 (1994)
59. Subramanian, N., Chung, L.: Metrics for software adaptability. In: Software Quality Management Conference, Loughborough, UK (2001)
60. Perez-Palacin, D., Mirandola, R., Merseguer, J.: On the relationships between QoS and software adaptability at the architectural level. J. Syst. Softw. **87**(1), 1–17 (2014)
61. Zhou, Z., Gaaloul, W., Gao, F., Shu, L., Tata, S.: Assessing the replaceability of service protocols in mediated service interactions. Future Gener. Comput. Syst. **29**(1), 287–299 (2013)
62. Wombacher, A., Li, C.: Alternative approaches for workflow similarity. In: 2010 International Conference on Services Computing, Miami, Florida, USA, July 2010, pp. 337–345
63. Dijkman, R.M., Dumas, M., García-Bañuelos, L.: Graph matching algorithms for business process model similarity search. In: 7th International Conference on Business Process Management, Ulm, Germany, September 2009, pp. 48–63
64. Dijkman, R.M., Dumas, M., van Dongen, B.F., Käärik, R., Mendling, J.: Similarity of business process models: metrics and evaluation. Inf. Syst. **36**(2), 498–516 (2011)
65. Kunze, M., Weidlich, M., Weske, M.: Behavioral similarity—a proper metric. In: 9th International Conference on Business Process Management, Clermont-Ferrand, France, August, September 2011, pp. 166–181
66. Fettke, P., Loos, P., Zwicker, J.: Business process reference models: survey and classification. In: Business Process Management Workshops, Nancy, France, September 2005, pp. 469–483
67. Lenhard, J., Schönberger, A., Wirtz, G.: Edit distance-based pattern support assessment of orchestration languages.: In: 19th International Conference on Cooperative Inforamtion Systems, Hersonissos, Crete, Greece, October 2011, pp. 137–154
68. Li, C., Reichert, M., Wombacher, A.: On measuring process model similarity based on high-level change operations. In: 27th International Conference on Conceptual Modeling, Barcelona, Spain, October 2008, pp. 248–264
69. Becker, M., Laue, R.: A comparative survey of business process similarity measures. comput. ind. **63**(2), 148–167 (2012)

70. Bae, J., Liu, J.C.L., Rouse, W.B.: Process mining, discovery, and integration using distance measures. In: International Conference on Web Services, Chicago, USA, September 2006, pp. 479–488
71. Zha, H., Wang, J., Wen, L., Wang, C., Sun, J.: A workflow net similarity measure based on transition adjancency relations. Comput. Ind. **61**(5), 463–471 (2010)
72. Weidlich, M., Mendling, J., Weske, M.: Efficient consistency measurement based on behavioral profiles of process models. IEEE Trans. Softw. Eng. **37**(3), 410–429 (2011)
73. Zakaria, Z., Atan, R., Ghani, A., Sani, N.: Unit testing approaches for BPEL: a systematic review. In: 16th Asia-Pacific Software Engineering Conference. Malaysia, Penang, December 2009, pp. 316–322
74. Lübke, D.: Unit testing BPEL Compositions. In: Test and Analysis of Service-oriented Systems, pp. 149–171. Springer. ISBN: 978-3540729112 (2007)
75. van der Aalst, W.M.P., Lohmann, N., Massuthe, P., Stahl, C., Wolf, K.: From public views to private views—correctness-by-design for services. In: 4th International Workshop on Web Services and Formal Methods, Australia, Brisbane, September 2007, pp. 139–153
76. van der Aalst, W.M.P., Dumas, M., Ouyang, C., Rozinat, A., Verbeek, E.: Conformance checking of service behavior. ACM Trans. Internet Technol. **8**(3), 29–59 (2008)
77. Both, A., Zimmermann, W.: Automatic protocol conformance checking of recursive and parallel BPEL systems. In: Sixth IEEE European Conference on Web Services, Dublin, Ireland, November 2008, pp. 81–91
78. García-Fanjul, J., Claudio de la Riva, J.T.: Generation of conformance test suites for compositions of web services using model checking. In: Testing: Academic and Industrial Conference—Practice and Research Techniques, United Kingdom, Windsor, August 2006, pp. 127–130
79. Cesari, L., Lapadula, A., Pugliese, R., Tiezzi, F.: A tool for rapid development of WS-BPEL applications. In: 25th ACM Symposium on Applied Computing, Switzerland, March, Sierre 2010, pp. 2438–2442
80. Simon, B., Goldschmidt, B., Kondorosi, K.: A human readable platform independent domain specific language for bpel. In: Second International Conference on Networked Digital Technologies, Prague, Czech Republic, July 2010, pp. 537–544
81. Hallwyl, T., Henglein, F., Hildebrandt, T.: A standard-driven implementation of WS-BPEL 2.0. In: 25th ACM Symposium on Applied Computing, Sierre, Switzerland, March 2010, pp. 2472–2476
82. González, L.S., Rubio, F.G., González, F.R., Velthuis, M.P.: Measurement in business processes: a systematic review. Bus. Process Manage. J. **16**(91), 114–134 (2010)
83. Cosmo, R.D., Vouillon, J.: On Software Component Co-Installability. In: ACM SIGSOFT Symposium on the Foundations of Software Engineering. Hungary, Szeged, September 2011, pp. 256–266
84. IETF: Metrics for the Evaluation of Congestion Control Mechanisms, March 2008, IETF Network Working Group
85. Zhou, Z., Bhiri, S., Zhuge, H., Hauswirth, M.: Assessing service protocol adaptability based on protocol reduction and graph search. Concurrency and Comput. Pract. Experience **23**(9), 880–904 (2011)
86. Karlsson, J., Wohlin, C., Regnell, B.: An evaluation of methods for prioritizing software requirements. Inf. Softw. Technol. **39**(14–15), 939–947 (1998)

Chapter 9
Business Process Intelligence Tools

Johannes Schobel and Manfred Reichert

Abstract This chapter analyzes contemporary software tools enabling Business Process Intelligence (BPI). BPI is one of the emerging trends in enterprise computing, which allows companies and organizations to maximize the value derived from their business processes. Moreover, BPI constitutes an umbrella term that summarizes different software tools, methods and best practices for real-time process analytics. The chapter presents an analysis of the features as well as the strengths and weaknesses of contemporary BPI tools along two characteristic application strategies of modern enterprises.

Keywords Business process intelligence · Process performance measurement · Process mining

9.1 Introduction

Business process intelligence (BPI) constitutes an umbrella term that summarizes different software tools, methods and best practices for real-time analytics, monitoring, decision support, and root-cause analysis in respect to operational business processes [4, 14].

The main goal of BPI is to either monitor individual business activities or entire business processes in order to reveal their malfunctions and inefficiencies. Based on respective analysis, new business strategies can be derived and optimized business process support can be provided for enterprises.

Currently, many BPI tools are available on the market. However, as each of these tools have different strengths and weeknesses, or provide specific unique features, it is often difficult to select the most suitable tool when introducing BPI in an enterprise.

J. Schobel (✉) · M. Reichert
Institute of Databases and Information Systems, Ulm University, Ulm, Germany
e-mail: johannes.schobel@uni-ulm.de

M. Reichert
e-mail: manfred.reichert@uni-ulm.de

© Springer International Publishing AG 2017
G. Grambow et al. (eds.), *Advances in Intelligent Process-Aware Information Systems*, Intelligent Systems Reference Library 123,
DOI 10.1007/978-3-319-52181-7_9

To provide a representative as well as detailed overview of contemporary BPI tools, we analyzed two strategies for their application in an enterprise. In order to achieve reproducible and valid results, we focused on a common method to evaluate these tools. Furthermore, all software tools were analyzed, tested and evaluated by BPI experts.

To be able to compare the features of the considered tools, we further set up a realistic test scenario. Moreover, a sophisticated data set from an operational SAP system was used and analyzed with the considered BPI tools in order to evaluate their features as well as their strengths and weaknesses.

9.1.1 Strategies for Using a BPI Tool

This section outlines two major strategies for using a BPI tool in an enterprise or organization. These strategies were elaborated in cooperation with BPI experts as well as consultants from the field of business intelligence. Both strategies provide typical scenarios for the use of BPI tools.

Stragety I (One Time Usage). In the context of a *one time usage* strategy, the BPI tool is solely used in a particular application (e.g., environment project) or for a restricted period of time. To foster such short-term usage, the BPI tool should be self-explanatory and easy to use. Moreover, it should offer simple data integration mechanisms. Usually, the analytical capabilities provided by a BPI tool of this category are not very powerful. Moreover, a quick overview on a specific business process or a set of activities is considered as most important. Thus, the goal is not to provide in-depth analysis.

Strategy II (Long Term Usage). The second strategy we consider is the *long term usage* of a BPI tool. Regarding this strategy, the BPI tool should be connected with all relevant information systems of the respective enterprise architecture. Usually, a multitude of predefined connectors to software systems, providing information on operational business processes, is available. The long term usage of a BPI tool enables profound insights into operational business processes, which fosters decision making and process optimization [11].

9.1.2 Dimensions for Evaluating BPI Tools

To evaluate contemporary BPI tools against the two strategies (cf. Sect. 9.1.1) we divided these scenarios into five different dimensions based on attributes applied in software selection schemes. However, certain dimensions are more important for certain strategies than others. As example consider dimension *Visualization*. On one hand, when applying a BPI tool solely to a single project, the visualization of discovered results is less important. On the other, when using BPI tools in the long term, the use of a variety of techniques enabling visual analytics allows for insights into existing business processes and their data.

The remainder of this chapter is structured as follows: Sect. 9.2 describes the methodology for analyzing BPI tools, whereas Sect. 9.3 presents a case study we conducted to evaluate these tools. In Sect. 9.4, assessment criteria and relevant attributes are described. Section 9.5 provides information about the BPI tools we evaluated. Section 9.6 presents detailed results. The chapter concludes with a discussion and summary in Sect. 9.7.

9.2 Methodology

In the context of the BPI tool study, we apply the *Analytic Hierarchy Process* (AHP) [15]. AHP constitutes a systematic procedure for representing the elements of a problem. In particular, it breaks the latter down into smaller components, and compares the latter pairwise to develop priorities at each level. Though AHP is unable to find the *correct* decision, it helps with identifying that one, which suits best to the problem at hand.

In general, AHP serves several purposes. On one hand, it allows finding a proper solution for a given problem, minimizing the time required for this. On the other, it provides comprehensible and reproducible results for decision making. Moreover, AHP fosters the discovery of inconsistencies as well as the comprehension of the problem.

The AHP process can be structured into several steps, which are easier to comprehend and deal with (compared to the overall process). Figure 9.1 visualizes these steps.

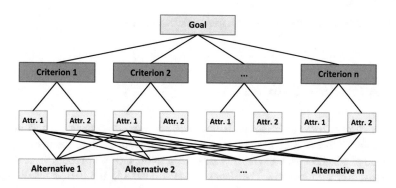

Fig. 9.1 AHP phase model

Table 9.1 AHP values for rating and comparing the attributes and dimensions

Value	Meaning	Description
1	Equal importance	Two activities equally contribute to the given objective
3	One alternative has a slightly higher importance	Experience and judgment slightly favor one activity over another
5	Essential or strong importance	Experience and judgment strongly favor one activity over another
7	Demonstrated importance	An activity is strongly favored and its dominance is practiced
9	Absolute importance	The evidence favoring one activity over another is the highest possible order of confirmation
2, 4, 6, 8	Intermediate values	Compromise (when needed)

1. First Step:

   - Define the problem and the knowledge needed.
   - Gather required data and criteria.
   - Structure the hierarchy from the top level (i.e., broad perspective on the problem) through intermediate levels (i.e., criteria) to the bottom level (i.e., alternatives).

2. Second Step:

   - Construct comparison matrices for each pair of criteria. Thereby, a particular criterion might dominate others. Furthermore, express these weightings as integers (cf. Table 9.1). This step is accomplished for all levels of the hierarchy.

3. Third Step:

   - Weight all criteria based on a mathematical model.

9.3 A Practical Case

As a benchmark for evaluating the BPI tools, a *purchase-to-pay* process, which had been extracted from a SAP system, was used. More precisely, the considered data set consisted of 519,633 events, 26,807 cases, 15 activities, 6,769 process variants, 344 resources, and 34 attributes. The data set was collected in the period from August 2011 to October 2011 (3 months). In order to enable a better understanding of the data and the respective business process, first of all, we present the process (cf. Fig. 9.2).

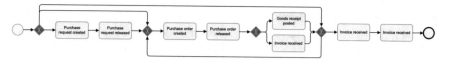

Fig. 9.2 The purchase-to-pay process (BPMN 2.0 notation)

9.3.1 Purchase-To-Pay Process

The purchase-to-pay process is one of the service processes of an enterprise's business. It describes the sequence of events and activities that starts with a demand for goods or services and ends with the payment. Most activities of this process can be characterized as transactional. Per definition, transactional activities are labour-intensive, but can be simply performed. However, as the purchase-to-pay process does not add any value to products or services of the enterprise, generally, it is considered as a cost driver.

The purchase-to-pay process looks simple at first glance: Starting with a demand for goods or services, a purchase request is created. Often, requests for quotations are then sent to potential vendors whose responses are rated against each other. The final purchase order contains information about the vendor for which the good or service is ordered, the quantity of items to be bought, and the related prices. Goods are delivered and related stock and inventory accounts are updated by receipt postings for the corresponding goods. Once services have been performed, a service acceptance is recorded through a separate workflow. In practice, however, there exist numerous deviations from this standard process, which move it from being simple and straightforward to become difficult and complex: Down payments might be involved, ordered goods be sent in more than one delivery, or goods be returned. Furthermore, purchase orders might not be created or be identifiable on the invoice. Moreover, invoiced prices might not match the prices of the purchase order, deficient goods receipts be recorded or invoices not be matched automatically (e.g., when they arrive prior to the goods).

9.3.2 Leveraging the Strengths of BPI Tools

Process complexity and the large amount of available information in existing information systems leverage the strengths of BPI tools [5]. Traditionally, purchase-to-pay analyses are difficult to perform for several reasons. First, three separate business departments with "silo thinking" are involved: purchase, logistics, and accounting. Second, the "to-be process" descriptions in enterprises do not precisely fit reality or "as-is process" descriptions are opaque. Finally, interviewees rarely have end-to-end information or provide only the minimum required information. Usually, process owners have a good understanding about tasks performed incorrectly as well as potential improvements. However, they lack accurate information to support their opinion. Thus, companies never gain transparency and full insights into the actual end-to-end process. Techniques provided by BPI tools, in turn, support decision-makers in gaining further insights into business processes and related operational data. For example, BPI tools allow users to drill down the "as-is process" and to visually identify deficiencies together with their root causes. In turn, this enables process owners as well as process experts to derive suitable counter measures. However, literature on practical applications of BPI techniques in the large scale is still scarce nowadays. Furthermore, there exist specific BPI tools on the market providing specific advantages and disadvantages.

We conducted an extensive analysis of available BPI tools. The main purpose of this tool analysis was to demonstrate the applicability of the provided techniques to real business needs as well as to determine the systems suited best for achieving process optimizations [22].

9.4 Assessment Criteria and Strategies

This section presents the assessment criteria we used to rate the considered BPI tools. Furthermore, strategies for using the latter are described.

9.4.1 Strategies

The study examines two different strategies for using a BPI tool. These strategies are *One Time Usage* of a BPI tool on one hand and its *Long Term Usage* on the other.

One Time Usage means that the tool is solely used for a particular project or a limited duration of time. Hence, a tool only requires simple connectors (e.g., CSV or Microsoft Excel) as well as fast and easy-to-use data integration facilities, as users will not be specifically trained to be able to use these tools. Regarding the one time usage of a BPI tool, providing a quick overview over the process or the data is therefore considered as most important.

Long Term Usage means that the BPI tool is intended to be used over a longer period of time. Therefore, it should be connected to all relevant information systems, which requires a variety of connectors (e.g., ERP connectors, database connectors, CSV or Microsoft Excel files). Through the integration of heterogeneous applications with the BPI tool, deeper insights into enterprise process data become possible (e.g., based on cross-process analyses) [9]. If a BPI tool is used for a longer period and a multitude of information systems need to be connected, higher installation, configuration and data integration efforts are acceptable.

9.4.2 Dimensions

The study considered five BPI tool dimensions: *system*, *data integration & extraction*, *data processing & analysis*, *mining*, and *data visualization*. The dimensions were elaborated in close consultation with domain experts to allow for a fine-grained classification. As AHP is used for evaluation purposes, every attribute has a specific weight within its corresponding dimension. Furthermore, each dimension is weighted depending on the strategy applied.

System This dimension describes how the tool behaves and how it can be handled. Attributes of this dimension include, for example, extensibility, usability and performance.

Data Integration & Extraction expresses how data can be extracted from enterprise information systems, providing data on the operational business processes (e.g., ERP systems, workflow management systems) [12, 22].

Data Processing & Analysis describes possibilities for analyzing the gathered data in the BPI tool. Common methods include Key Performance Indicators (KPIs) and Bottleneck Analysis.

Mining deals with extracting knowledge from a given data set (e.g., a database of event log) and transforming it into understandable structures such as process graphs or organization models [8].

Data Visualization examines the different ways of representing the results from data analyses or mining algorithms.

9.4.3 Attributes

Table 9.2 provides selected lists of attributes for the described dimensions. For a table comprising all attributes we refer to Appendix A.

9.5 Business Process Intelligence Tools

This chapter provides information about the BPI tools we assessed. We assigned the tools to different categories according to their origin. Examples of these categories include Business Intelligence and Business Process Management suites. All tools were rated in the same way and tested against the introduced strategies (cf. Sect. 9.1) based on the described data set (cf. Sect. 9.3).

Table 9.2 Selected list of assessment attributes

Dimension	Attribute	Description
System	Extensibility	Is it possible to extend the tool (e.g., based on available plug-ins or a provided application programming interface (API))?
	User and Rights Management	Refers to the ability to manage users as well as to define their permissions for importing, analyzing and visualizing data with the BPI tool
Data Integration & Extraction	Connector Handling	Refers to the intuitiveness of the connectors: are wizards provided guiding users through the process of data integration?
	Real-Time Measurement	Refers to the ability to integrate data in real-time based on push mechanisms
Data Processing & Analysis	Key Performance Indicators	Refers to the ability to define Key Performance Indicators (KPIs) with a formula editor. How powerful is this editor?
	Bottleneck Analysis	Is there a way to automatically identify bottlenecks in a business process?
Mining	Process Discovery	Refers to the ability to extract the as-is process from a given data set
	Conformance Checking	Compare the as-is process with the model of the to-be process and check for outliers and deviations
Data Visualization	Process Graph	Display results provided by process mining algorithms in a graph-like representation (e.g., in terms of a process model)
	Reports	Create a user-defined report and use it as a template in another project

9.5.1 Tool Shortlist

At the beginning of the tool study, relevant software vendors as well as their BPI tools were determined. The resulting longlist provided the basis for the following tool analysis and evaluation. To set a focus as well as to enable a hands-on assessment

Table 9.3 Shortlist of the evaluated BPI tools

Software vendor	System	Origin
Software AG	ARIS MashZone [16]	Business Process Intelligence
Software AG	ARIS PPM [17]	Business Process Performance Manager
Celonis GmbH	Celonis discovery [2]	Business Process Intelligence
Fluxicon	Disco & ProM [3]	Business Process Mining
Microsoft	Business intelligence [10]	Business Intelligence
QlikTech	QlikView [13]	Business Intelligence
TIBCO Software Inc.	TIBCO Spotfire [19]	Business Intelligence

of selected tools, the longlist was reduced to a shortlist that finally comprised seven BPI tools (cf. Table 9.3).

9.5.2   Tool Categories

Traditional Business Intelligence (BI) Tools BI [1] defines procedures and methods for the systematic analysis (i.e., collection, evaluation and visualization) of business data. The overall aim is to gain information that facilitates strategic as well as operational decisions taking defined business goals into account. Usually, this is accomplished based on historic data.

Process Mining Tools Process mining tools [7, 20] focus on the analysis of executed business processes. This is accomplished by analyzing event logs recorded by the information systems. The aim is to extract relevant knowledge from these logs; e.g., in order to optimize business processes. Process mining tools allow discovering process models verifying that given process model complies with imposed rules (i.e., constraints).

Business Process Intelligence Tools BPI tools focus on the real-time analysis of operational business processes. Compared to traditional BI tools, BPI tools are more focused on the analysis of operational business processes. Moreover, BPI tools not only provide simple KPIs, as common BI tools, but also detailed information about control flow, (social) interactions between the actors involved in the process, related documents (e.g., invoice documents), and relevant attributes (e.g., overall duration of the business process or effective working time regarding specific activities [6]).

Table 9.4 ARIS MashZone

Value	Description
+	Easy creation of business dashboards
+	Direct connection to ARIS Process Performance Manager
+	Application runs within a web browser
+	Convenient combination of data feeds with special editor
−	No process mining support
−	Missing support of more advanced process analyses
−	Limited support for defining KPIs
*	Intuitive approach for linking data and calculating KPIs
*	Easy data integration via drag & drop techniques
*	Easy data manipulation and transformation with wizards and editors

9.5.3   Fact Sheets for the Evaluated BPI Tools

This section presents key facts for each BPI tool we evaluated. The corresponding fact sheet includes a short description of the tool as well as the pros (marked with +) and cons (marked with −) discovered during its evaluation. Furthermore, specific features (marked with *) are elaborated. Note that the latter could not be compared with the other tools as they are unique for the respective tools.

ARIS MashZone ARIS MashZone is a free BI tool provided by Software AG [16]. It enables users to create and manage business dashboards. Since ARIS MashZone can run in a web browser, it may be accessed from anywhere. Table 9.4 summarizes the pros and cons of this tool as well as its unique properties.

ARIS Process Performance Manager ARIS Process Performance Manager is a BPI tool provided by Software AG [17]. It allows calculating Key Performance Indicators (KPIs) as well as visualizing the business processes of an enterprise. In particular, it enables comprehensive process analysis supports of different kind. Table 9.5 summarizes the pros and cons of this tool as well as its unique properties.

Celonis Discovery Celonis Discovery is a BPI tool developed by Celonis GmbH [2]. It focuses on the extraction of process knowledge from ERP systems (e.g., SAP) as well as on process optimization. Furthermore, Celonis Discovery may be used for obtaining a quick overview on the KPIs of the discovered processes. Since Celonis Discovery is able to run in a web browser, it may be accessed from anywhere. Table 9.6 summarizes the pros and cons of this tool as well as its unique properties.

Table 9.5 ARIS process performance manager

Value	Description
+	Visualizes process as graphs (i.e., EPCs)
+	Allows for the direct import of data from SAP systems
+	Provides an intuitive user interface for creating and managing dashboards
−	Very complex data integration
−	High configuration and customization efforts, resulting in an awkward handling of the Customization Tool Kit (CTK)
−	Cumbersome definition of performance indicators (KPIs)
*	Direct connection to SAP systems
*	Ability to mine processes across different systems

Table 9.6 Celonis discovery

Value	Description
+	Allows for an easy Microsoft Office integration to foster evaluation and analysis
+	Complete web application; i.e., runs in a web browser
+	Calculates KPIs for activities in a process model
−	Limited data import support (relies on third-party application)
−	Insufficient and cumbersome user management
−	Cumbersome editor for defining KPIs
*	Office integration for Microsoft Excel
*	Proprietary mining algorithms
*	KPIs for activities and state transitions

Disco & ProM Disco and ProM constitute process mining tools [21]. ProM constitutes an open source system released by TU Eindhoven, whereas Disco is a commercial tool developed by Fluxicon [3]. Disco focuses on data integration, data transformation, and data analysis based on process discovery techniques, whereas ProM provides a larger variety of algorithms that allow for detailed process analyses based on event logs. Table 9.7 summarizes the pros and cons of these tools as well as their unique properties.

Table 9.7 Disco/ProM

Value	Description
+	ProM comprises a multitude of mining algorithms
+	Very intuitive handling of Disco
+	Scientific environment and connections to research institutes
+	Provides a very quick overview of business processes
−	Time consuming integration of data in ProM
−	Disco provides an easy-to-use data transformation interface
*	Animation of business process execution
*	Displaying variants of business processes, highlighting their differences

Table 9.8 Microsoft business intelligence

Value	Description
+	Good performance when analyzing large amounts of data
+	Variety of formulas for representing KPIs
+	Distribution of KPIs and dashboards using Microsoft SharePoint Server
+	Comes free with Microsoft SQL Server
−	Very complex data integration
−	Missing process analysis features
*	Tight integration with Microsoft SharePoint Server and Microsoft Office
*	Integrated script editor (Plug-In for Microsoft Visual Studio) for managing the data cubes
*	Predefined data mining algorithms enabling sophisticated data analyses

Microsoft Business Intelligence Microsoft Business Intelligence is shipped free with the Microsoft SQL Server [10]. It enables the setup of data cubes and allows for a tight integration with Microsoft SharePoint Server. In turn, the latter enables users to distribute and visualize KPIs in an enterprise portal. Table 9.8 summarizes the pros and cons of this tool as well as its unique properties.

Table 9.9 QlikView

Value	Description
+	Fast and intuitive creation of business dashboards
+	Integrated script editor that allows extending the predefined functionality
+	Analyzing and visualizing data on mobile devices
+	Creating reports using drag & drop and publishing them on different channels
−	No process mining techniques available
−	No support for analyzing business processes
*	Support of using smart mobile devices for analyzing business data
*	Publication of business dashboards on web platforms (e.g., web page, portal)
*	Comes with a built-in programming language to enhance the system with user-defined functions

Table 9.10 TIBCO spotfire

Value	Description
+	User-friendly user interface
+	Easy-to-use data import function based on copy & paste
+	Intuitive creation of business dashboards
−	No process mining techniques available
−	No support for business process analyses
*	Commercial add-ons available to extend standard functionality
*	Integration of smart mobile devices for analyzing business data

QlikView QlikView is a BI tool distributed by QlikTech [13]. It allows for the easy creation of sophisticated business dashboards based on a powerful drag & drop editor. In addition, common data interfaces are supported to foster ease of use. Table 9.9 summarizes the pros and cons of this tool as well as its unique properties.

TIBCO Spotfire TIBCO Spotfire is a BI tool distributed by TIBCO Software [19]. Like QlikView, TIBCO Spotfire allows creating business dashboards quickly based on drag & drop techniques. Overall, TIBCO Spotfire allows for quick overviews on business data. Table 9.10 summarizes the pros and cons of this tool as well as its unique properties.

9.6 Applying the Tools to the Case

The bar charts presented in this section give insights into the experiences gathered when applying the process intelligence tools to the described sample process. The evaluation was carried out by experts from the areas of business (process) intelligence and process mining respectively. The data set (cf. Sect. 9.3) was imported by the respective tools and used to analyze process instances. Results were then normalized based on the given maximum values. In certain cases, the systems were not rated as 100% as the experts agreed that respective attributes could still be improved.

9.6.1 System

This section presents evaluation results regarding the *system* dimension (cf. Table 9.2); i.e., tool behaviour and tool handling. Figure 9.3 assesses the extensibility of the tools, whereas Fig. 9.4 evaluates the user & rights management in the respective tools. For example, Disco/ProM is rated by far best regarding extensibility, whereas Microsoft BI is rated best in respect to user & rights management.

Fig. 9.3 Results for attribute *extensibility*

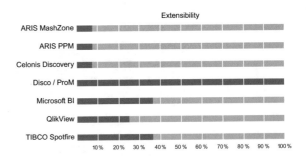

Fig. 9.4 Results for attribute *user and rights management*

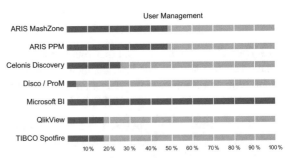

Fig. 9.5 Results for attribute *connector handling*

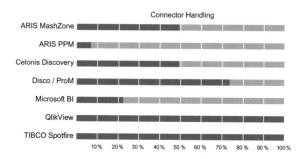

Fig. 9.6 Results for attribute *real time measurement*

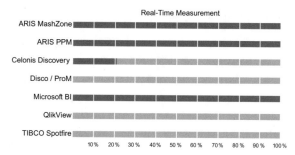

9.6.2   Data Integration & Extraction

This section presents evaluation results regarding the *data integration & extraction* dimension (cf. Table 9.2); i.e., the capabilities to import data from different sources (i.e., systems). Figure 9.5 shows the results in respect to connector handling, whereas Fig. 9.6 reveals the results for analyzing data in real time. As can be seen, QlikView and TIBCO Spotfire show the best results regarding the handling of data connectors as they provide an easy to use approach for integrating required data. In turn, the ARIS tools and Microsoft BI enable the monitoring of the respective data in real-time.

9.6.3   Data Processing & Analysis

This section presents evaluation results regarding the *data processing & analysis* dimension (cf. Table 9.2); i.e., the capabilities to process the extracted data within the BPI tool. Figure 9.7 presents the capabilities of defining and evaluating KPIs, whereas Fig. 9.8 presents bottleneck analysis capabilities for the analyzed tools. While Microsoft BI, Qlikview and TIBCO Spotfire are on a par, Disco & ProM is far behind. However, Disco & ProM is the only tool that provided sufficient capabilities for bottleneck analysis.

Fig. 9.7 Results for attribute *KPI*

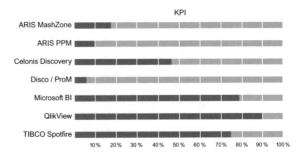

Fig. 9.8 Results for attribute *bottleneck analysis*

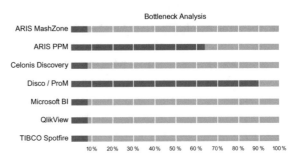

Fig. 9.9 Results for attribute *process discovery*

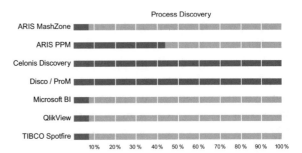

9.6.4 Mining

This section presents results for the *mining* dimension (cf. Table 9.2); i.e., the support of process discovery (cf. Fig. 9.9) and conformance checking (cf. Fig. 9.10). The tools originating from process mining tools (i.e., Disco, ProM, and Celonis Discovery) showed the best performance. Note, that several tools did not offer conformance checking features.

Fig. 9.10 Results for attribute *conformance checking*

Fig. 9.11 Results for attribute *process graph*

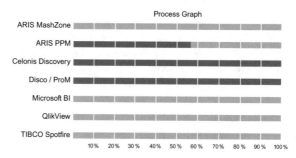

Fig. 9.12 Results for attribute *reports*

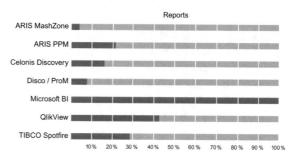

9.6.5 Data Visualization

This section provides results for dimension *data visualization* (cf. Table 9.2); i.e., the capabilities to present the processed data extracted from different sources. Figure 9.11 shows the features to represent mined processes as graphical process models (in common notations like BPMN 2.0 or EPCs). Furthermore, Fig. 9.12 presents the results we obtained when creating custom reports for the extracted and processed data. Obviously, the tools performing best in respect to attribute *process discovery* also perform best in respect to attribute *process graph* within this dimension. Microsoft BI is the most powerful tool regarding the creation of reports due to its tight integration with SharePoint and Office.

9.7   Discussion

Based on the results of the evaluation presented in Sect. 9.6, the following recommendations can be made regarding the use of business (process) intelligence systems.
In particular, these recommendations will foster system selection depending on the
given application environment and strategy respectively.

9.7.1   One Time Usage

As demonstrated in the graph below, it is not possible to nominate a clear winner as
none of the evaluated tools performs best in all dimensions. However, certain dimensions should be considered as more important than others. Figure 9.13 provides an
overall rating regarding the One Time Usage of BPI tools.

System In general, dimension *system* is not as important as others regarding the One
Time Usage strategy. The only attribute relevant in this context is usability. Regarding
the latter, Disco & ProM, QlikView, and TIBCO Spotfire obtain the best evaluation
results.

Data Integration & Extraction This dimension is crucial for any One Time Usage
strategy. Note that an easy and fast data integration is indispensable when analyzing business processes and their underlying data. For a simple BI use case, ARIS
MashZone and QlikView are the most suitable tools since they allow for a quick and
intuitive data integration. If a more powerful BPI tool is needed, the combined use
of Disco & ProM might be the right choice. In particular, both tools offers a variety

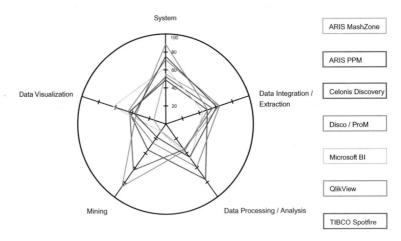

Fig. 9.13 One time usage—overall rating

of techniques for process analysis as well as for an intuitive and user-friendly data integration.

Data Processing & Analysis For a project-based strategy, data processing and data analysis are less important. Note that SixSigma Formulas [18] or user-defined KPIs are not always required to gain a broad overview on business processes. No tool clearly stands out in this dimension, i.e., all evaluated tools obtained middle-range scores.

Mining has a high priority as the purpose of the One Time Usage strategy is to gain a rough overview of the business processes. Furthermore, visualizing the business process reveals additional insights into the process itself. This dimension includes attributes like process discovery, conformance checking, and process visualization. The combination of Disco and ProM can be considered as winner since it provides a wide range of mining algorithms.

Data Visualization It is crucial to be able to visualize business data in an intuitive and quick way when applying the One Time Usage strategy. In general, however, it is not necessarily required to provide a wide range of visualization features (e.g., several different diagram types). TIBCO Spotfire and QlikView are comprehensively convincing providing powerful visualization possibilities and easy-to-use diagrams.

> One Time Usage Recommendation
>
> Altogether, it becomes necessary to distinguish the strategy depending on how the tool is going to be used. If a BPI tool is needed, the combination of Disco & ProM is recommended in the context of the One Time Usage strategy. If a BI tool is required, QlikView or TIBCO Spotfire allow for a fast and quick overview on the data gathered.

9.7.2 Long Term Usage

Like for the One Time Usage strategy, it is not possible to define a clear winner for the Long Term Usage strategy. None of the evaluated tools clearly wins or loses the evaluation. Nevertheless, regarding the Long Term Usage strategy, we identified some differences regarding the importance of the evaluated dimensions. Figure 9.14 shows the overall rating for the Long Term Usage of BPI tools.

System In general, this dimension is not as important for Long Term Usage as the others. The only attribute, which may be considered as important, refers to the usability of the tool. Regarding this attribute, Disco & ProM, QlikView, and TIBCO Spotfire perform best.

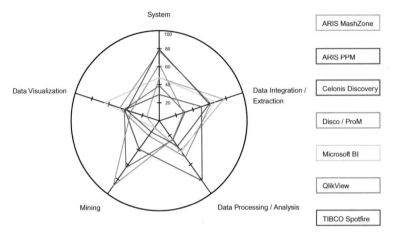

Fig. 9.14 Long term usage—overall rating

Data Integration & Extraction This dimension is rated lower for the Long Term Usage strategy since a more complex data integration pays off for it. Attributes like real-time measurement and providing predefined connectors are more important compared to any One Time Usage strategy. Overall, ARIS PPM and Microsoft BI achieved the highest rating since both allow retrieving data from several systems in real time.

Data Processing & Analysis is more important for Long Term Usage than for One Time Usage of the tool. Attributes such as benchmarking or defining KPIs are required for enabling an in-depth analysis of the data set. If a BPI tool with a variety of data processing algorithms is needed, ARIS PPM can be used. However, its usability is not as intuitive as the one of other tools evaluated. Instead, Celonis Discovery can be considered as a better and more user-friendly choice if fewer possibilities to analyze enterprise data are sufficient.

Mining constitutes an important dimension in the context of the Long Term Usage strategy. It includes attributes like process discovery or conformance checking. The combination of Disco and ProM performs best in this dimension due to the variety of mining algorithms provided and process modeling notations supported.

Data Visualization It is useful to visualize data in different ways for a Long Term Usage of the BPI tool (e.g., dashboards or reports). Microsoft BI allows to publish results via Microsoft SharePoint Server, which constitutes an advantage compared to all other tools we evaluated.

Fig. 9.15 Overall results for
one time usage

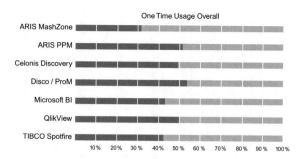

Fig. 9.16 Overall results for
long term usage

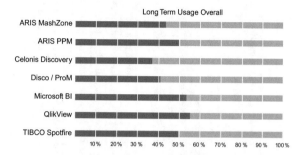

> **Long Term Usage Recommendation**
>
> If an intense initial setup and configuration phase is acceptable, ARIS
> PPM offers a variety of different features for analyzing and processing
> data. If this does not apply, Celonis Discovery or the combined use
> of Disco and ProM are alternatives. Microsoft Business Intelligence is
> recommended for the Long Term Usage of a BI tool due to its tight
> integration with Microsoft SharePoint Server, which allows realizing a
> company portal enriched with business dashboards.

The overall results, both for the One Time and Long Term Usage of BPI tools are
provided in Figs. 9.15 and 9.16.

A Assessment Attributes

See Table 9.11.

Table 9.11 Complete list of all assessment attributes

	Attribute	Description
System	Extensibility	Is it possible to extend the system (e.g., based on plug-ins or an API)? Is the system open source?
	Usability	Refers to the ease of use and intuitiveness of the system. This includes learnability, efficiency, memorability and satisfaction
	Performance	Refers to the system's ability to handle small (<1 million instances), medium, or large (>10 million instances) data sets in reasonable time
	User & Rights Management	Refers to the possibility to manage users as well as to set up their rights and permissions to import, analyze and visualize data within the system
Data Integration & Extraction	Connector Handling	Refers to the intuitiveness of handling connectors. Are wizards available guiding the user through the process of data integration?
	Connectors	Which connectors are available (e.g., connectors for SAP, CSV, XLS, or DBMS)?
	Upload Interface	How easy is it to import external data (e.g., field assistant, wizards)?
	Real-Time Measurement	Refers to the ability to integrate data in real-time base on push mechanisms
	One Time Active Extraction	Refers to the ability to integrate data once based on pull mechanisms. Is the system able to import data from a given source?
	Continuous Active Extraction	Refers to the ability to integrate data based on continuous pull mechanism from a source
	Process Template	Are there predefined data integration templates for common source systems (e.g., SAP)?
	Customized Templates	Is it possible to create user-defined templates for data integration and to reuse them in other projects?
Data Processing & Analysis	Benchmarking	Compare "as-is" and "to-be" states. How many processes can be compared at once?
	KPI	Refers to the ability to define KPIs with a specific formula editor. How powerful is this editor?
	SixSigma Formulas	Are predefined SixSigma formulas available and ready-to-use?
	Ad-Hoc Analysis	Does the system support ad-hoc analyses (e.g., filter attributes)?
	Bottleneck Analysis	Is there a way to automatically identify bottlenecks for a given business process?
	Alerting & Compliance Monitoring	Is there a way to define business rules and to assign alerts when these rules are violated?
	Extraction & Export	Refers to the ability to export data in different formats for their further processing within third party systems (e.g., Microsoft Excel, Visio, PDF)
	Saving of Analysis	Is it possible to save ad-hoc analysis, conserve the status of a dashboard or a chart, and bookmark frequently used queries?

(continued)

Table 9.11 (continued)

	Attribute	Description
Mining	Data Mining	Does the system allow for the use of data mining algorithms (e.g., k-means algorithm)?
	Social Mining	Refers to the mining of communication and organizational structures within a process (e.g., handover of work, social network analysis)
	Process Discovery	Refers to the ability to extract the "as-is" process from a given event log
	Process Visualization	Refers to the visualization of business processes using a process modelling language (e.g., EPC or BPMN)
	Filter Possibilities	Refers to the ability to filter data on important attributes (e.g., location, fiscal year) using drill-down and roll-up features
	Conformance Checking	Compare the "as-is process" with the "to-be process" and check for outliers and violations
	Variant Mining	Determine different variants of a business process. How relevant is a particular process variant compared to others?
Data Visualization	Table	Display analysis results as table
	Balanced Scorecards	Display analysis results as a balanced scorecard
	Process Graph	Display results provided by process mining algorithms as graph (e.g., process model)
	Dashboards	Combine different visualizations in a dashboard (e.g., table, diagrams and graphs)
	Reports	Create a user-defined report and reuse it as template
	Publication	Distribute and publish the data on different channels (e.g., website, PowerPoint presentation)
	Customization	Customize the visualization of analyses results to comply with the corporate design of the company
	Interactive Filtering/Drill-Down	Filter diagrams or other visualization elements interactively

References

1. Anandarajan, M., Anandarajan, A., Srinivasan, C.A.: Business intelligence techniques: a perspective from accounting and finance. Springer Science & Business Media (2012)
2. Celonis GmbH: Celonis Process Mining. http://www.celonis.de/. Accessed 03 Mar 2015
3. Fluxicon: Process Mining and Process Analysis—Fluxicon. http://fluxicon.com/disco/. Accessed 03 Mar 2015
4. Grossmann, W., Rinderle-Ma, S.: Fundamentals of Business Intelligence. Springer (2015)
5. Hipp, M., Michelberger, B., Mutschler, B., Reichert, M.: Navigating in process model repositories and enterprise process information. In: IEEE 8th International Conference on Research Challenges in Information Science (RCIS 2014), pp. 1–12. IEEE Computer Society Press (2014)
6. Lanz, A., Weber, B., Reichert, M.: Time patterns for process-aware information systems. Requirements Eng. **19**(2), 113–141 (2014)
7. Li, C., Reichert, M., Wombacher, A.: The MinAdept clustering approach for discovering reference process models out of process variants. Int. J. Cooper. Inf. Syst. **19**(3 & 4), 159–203 (2010)
8. Ly, L.T., Rinderle, S., Dadam, P., Reichert, M.: Mining staff assignment rules from event-based data. In: Proceedings of Workshop on Business Process Intelligence (BPI) in conjunction with (BPM'05), pp. 177–190. No. 3812 in LNCS. Springer (2005)
9. Michelberger, B., Mutschler, B., Reichert, M.: Towards process-oriented information logistics: why quality dimensions of process information matter. In: Proceedings of 4th International Workshop on Enterprise Modelling and Information Systems Architectures (EMISA 2011), pp. 107–120. No. 190 in Lecture Notes in Informatics (LNI). Koellen-Verlag (2011)
10. Microsoft AG: Business Intelligence in Office and SQL Server—Microsoft. www.microsoft.com/en-us/server-cloud/solutions/business-intelligence/. Accessed 03 Mar 2015
11. Mutschler, B., Reichert, M., Bumiller, J.: Unleashing the effectiveness of process-oriented information systems: problem analysis, critical success factors, and implications. IEEE Trans. Syst. Man Cybernet. Part C: Appl. Rev. **38**(3), 280–291 (2008)
12. Mutschler, B., Weber, B., Reichert, M.: Workflow management versus case handling: results from a controlled software experiment. In: 23rd Annual ACM Symposium on Applied Computing (SAC'08), Special Track on Coordination Models, Languages and Architectures, pp. 82–89. ACM Press (2008)
13. QlikTech: Analytics, Data Discovery, Data Visualization, QlikView BI Dashbaord|Qlik. http://www.qlik.com/en/explore/products/qlikview. Accessed 03 MAr 2015
14. Reichert, M., Weber, B.: Enabling Flexibility in Process-Aware Information Systems: Challenges, Methods, Technologies. Springer, Berlin (2012)
15. Saaty, T.L., Vargas, L.G.: Models, Methods, Concepts and Applications of the Analytic Hierarchy Process, vol. 175. Springer Science & Business Media (2012)
16. Software AG: ARIS MashZone—Business Mashups. http://www.mashzone.com/. Accessed 03 Mar 2015
17. Software AG: Software AG Process Performance Manager. http://www.softwareag.com/corporate/products/apama_webmethods/intelligence/products/process_performance/overview/default.asp. Accessed 03 Mar 2015
18. Tennant, G.: Six Sigma: SPC and TQM in Manufacturing and Services. Gower Publishing Ltd., Aldershot (2001)
19. TIBCO Software Inc.: TIBCO Spotfire—Business Intelligence Analytics Software & Data Visualization. http://spotfire.tibco.com/. Accessed 03 Mar 2015
20. Van Der Aalst, W.: Process Mining: Discovery, Conformance and Enhancement of Business Processes. Springer Science & Business Media (2011)

21. Van Der Aalst, W., Adriansyah, A., de Medeiros, A.K.A., Arcieri, F., Baier, T., Blickle, T., Bose, J.C., van den Brand, P., Brandtjen, R., Buijs, J., et al.: Process mining manifesto. In: Business Process Management Workshops, pp. 169–194. Springer (2012)
22. Weber, B., Mutschler, B., Reichert, M.: Investigating the effort of using business process management technology: results from a controlled experiment. Sci. Comput. Program. **75**(5), 292–310 (2010)

Printed in the United States
By Bookmasters